HERDING HEMINGWAY'S CATS

Also available in the Bloomsbury Sigma series:

HERDING
HEMINGWAY'S
CATS

UNDERSTANDING HOW
OUR GENES WORK

Kat Arney

BLOOMSBURY
sigma

Bloomsbury Sigma
An imprint of Bloomsbury Publishing Plc

50 Bedford Square
London
WC1B 3DP
UK

1385 Broadway
New York
NY 10018
USA

www.bloomsbury.com

BLOOMSBURY and the Diana logo are trademarks of
Bloomsbury Publishing Plc

First published 2016

Quote on p. 63 used with permission of Arla Foods.
Quote on pp. 69–70 reprinted by permission of Edward Monkton.

British Library Cataloguing-in-Publication Data
A catalogue record for this book is available from the British Library.

Library of Congress Cataloguing-in-Publication data has been applied for.

ISBN (hardback) 978-1-4729-1004-2
ISBN (trade paperback) 978-1-4729-1005-9
ISBN (ebook) 978-1-4729-1006-6

2 4 6 8 10 9 7 5 3 1

Typeset in Bembo Std by Deanta Global Publishing Services,
Chennai, India
Printed and bound in Great Britain by CPI Group (UK) Ltd,
Croydon CR0 4YY

Bloomsbury Sigma, Book Nine

To find out more about our authors and books visit www.bloomsbury.com.
Here you will find extracts, author interviews, details of forthcoming events and
the option to sign up for our newsletters

To Mum and Dad, with much love
Thanks for the nature and the nurture

Contents

It's All About That Base

It all started with a photo of a cat. I was hiding at the back of a scientific conference at the Royal Society in London when a cuddly looking cat with unusually big feet caught my eye. 'This is a Hemingway cat,' said the lecturer, pointing at the animal on the large screen behind him. 'They have six toes – they're polydactyl. Ernest Hemingway was said to be fond of them, and they still live on his estate in Florida today. And here …' he poked at the computer, changing the slide to one covered with photos of misshapen human hands, '… are polydactyl children with extra digits. It's the same genetic mistake that causes them.'

Looking at a six-toed cat or six-fingered human, a natural assumption might be that it's due to a fault in a gene. But it's not. In fact the cause lies in a faulty region of DNA that acts as a control switch, normally turning a gene on at the right time in the right place to direct the formation of fingers and toes as a baby or kitten grows in the womb. Not only that, but the switch is miles away (in molecular terms) from the gene it acts upon. Learning about the Hemingway cats and their broken switches got me thinking about my own understanding of how genes work, and how I explain it to the public through my work as a science writer and broadcaster.

❁

My first real brush with modern genetics came while I was at secondary school, courtesy of our formidable deputy headmaster Mr Myers. As well as stalking the school corridors with a steely glare of stern disapproval, doling out detentions seemingly at random, he also doubled as a biology teacher. One day it was his turn to preside over the regular school assembly. We dutifully trooped into the main hall to sit cross-legged on the floor, doing our best to avoid his eye. He took to the stage, black academic gown flowing out behind him like a cape, clasping in his hand what looked like a magazine but must have been a scientific journal of some kind. Towering in impotent fury from the stage, he shook it at us in disapproval as if it were a piece of pornography fished out from behind a cistern in the boys' toilets. 'Look at this!' he thundered, slapping at a page covered in the letters A, C, T and G, repeated in seemingly endless permutations. 'It's like the phone book! All these letters. Letters, letters, letters.' A pause for breath. 'THIS IS BIOLOGY NOWADAYS!'

While Mr Myers may have believed that ACTG was a four-letter word, the language of genes has made its way into common parlance over recent years. Genes are the things in your DNA that make your eyes blue, your belly bulge or your hair curl. The newspapers tell us that they control our risk of cancer, heart disease, alcoholism, Alzheimer's and more. A thousand dollars will buy you your very own genomic readout, sequenced in a matter of days and neatly stored on a USB stick. Genetic knowledge has the power to save us – there are drugs targeting the products of faulty genes in tumours, and recent advances in gene therapy that let the blind see. And we can trace the march of genes writ large in the seemingly endless variations of life on the face of our planet – evolution is just genetics plus time, after all.

Today we merrily talk about all sorts of things being 'in our genes', from a talent for singing to a life-threatening cancer. My mother is obsessed with family history, believing that almost every aspect of mine and my sisters' characters can be traced back to one or other of my long-dead ancestors, be they Baptist preachers or depressive alcoholics.* Science writers enthusiastically describe the genome – the sum total of an organism's DNA – as the blueprint of life, akin to a computer program or architect's plan. The double helix has become a cultural icon, not to mention lazy advertising

---

*Turns out I got both. THANKS, MUM.

shorthand for 'Ooh, science!' But while the language of genetics has infiltrated the public consciousness, a genuine understanding of what our genes are and what they do has not.

Most biology textbooks define genes as particular strings of DNA 'letters' – chemicals known as bases – and there are four of them in common use in all living things: A, C, T and G, short for adenine, cytosine, thymine and guanine. The particular order of the letters encodes instructions telling the cells in your body to make various molecules, in the same way that different recipes encode the directions for making cakes, pies or stews.

Unfortunately, it's not as simple as viewing our genome as a biological Mrs Beeton or Martha Stewart. As technology develops and scientists discover more about the secrets within our cells, the picture has begun to get very murky indeed. More than 2 metres (6½ feet) of DNA is packed into almost every cell of your body, crammed with thousands of genes that need to be turned on and off at the right time and in the right place. Rather than a neatly bound set of recipes, the genome as we understand it today is a dynamic, writhing library, buzzing with biological readers and writers. The text is constantly copied, tweaked and occasionally even torn up altogether. Every volume bulges with annotations and sticky notes, and there are thousands of pages that just seem to be complete nonsense. The cataloguing system would give even the hardiest librarian a nervous breakdown. Yet out of this chaos, we create life.

Right from the start, when a lucky sperm enters an egg, we're built by a complex biological ballet of gene activity enabling us to grow from a single cell into a baby. Intricate structures form: the twisting tubes of a heart, the rippling folds of a brain, the birdlike bones of a tiny skeleton. Along the way, each type of cell develops its own specialism. Skin cells form sturdy, neat layers to keep our insides in and the outside out, while our brain cells crackle with tiny pores transmitting electrical impulses. All of this is controlled by our genes. But how are they doing it? *How do our genes really work?*

In search of answers, I travelled and Skyped my way around the world talking to a range of researchers working at the frontiers of genetics, past and present. There wasn't any particularly clever selection process behind my hit list. In some cases they were people I already knew from my previous incarnation as a research scientist. Others were recommended to me, or just seemed like interesting people to go and see, based on the ideas and work they had published. Not all of their stories are told here, but every single person I spoke to shaped my thinking in some way. I'm grateful to them all for their time, intellectual generosity and coffee. Lots of coffee.

As well as asking everyone about their research, I always ended each interview with the same question: What do you think is weird? The history of science tells us that today's inexplicable oddities are tomorrow's revolutionary break-throughs, and I wanted to look down the road we're heading along, as well as mapping out our current understanding. When I started working on this book I thought I was pretty clued up. After all, I'd spent years studying genetics and working in research labs dedicated to figuring out what's going on as our genes are read and interpreted. But as I started talking, reading and thinking about it all, my preconceptions skittered away from me like wilful kittens. It turns out that there's a lot less certainty – and a lot more myth and dogma – around our genes than many people might realise.

This is my attempt to herd all these ideas together. It's not a hardcore genetics textbook, by any means. Instead, it's a book full of stories about you, your genes, cats with thumbs, fish with hips, and some of the wonderful ways in which your genome comes to life to make you who you are. I hope you enjoy it.

🐾

A couple of things to note: I use the words 'read' and 'reading' in two different contexts throughout. One is the act of scientists reading the order of letters (bases) in DNA – a technique known as DNA sequencing. The other refers to the machinery inside a cell reading a gene when it is switched on, making a copy of the information within it in the form of a message called RNA. This process is more formally known as transcription, and I will talk about genes being transcribed or read into RNA. Hopefully it should be obvious which context is relevant in each case.

I tend to use the word 'letters' to describe A, C, T and G – the four bases in RNA and DNA that make up the alphabet of life. In RNA the letter T is replaced by a very similar base called U. For simplicity's sake I've stuck with T throughout the book for both RNA and DNA. I've also avoided using standard nomenclature for gene names to avoid cluttering the pages with arcane capitalised acronyms.

There are plenty of excellent books that go into more depth about the mechanisms of life if you want to know more, and I've mentioned some of my favourites in the footnotes and references. Also at the back you'll find links, research papers and further reading, along with a handy glossary of technical terms.

CHAPTER ONE

It's Not What You've Got, It's What You Do With It That Counts

There is no God, there is no free will, there's just DNA.
Kari Stefansson, DeCODE founder

Eric Miska is not a happy man. 'It's very complicated,' he sighs. 'We have all this data, but we have no functional understanding. On the David Letterman TV show they used to have this section called "Is this anything?" and the audience had to vote on whether something is interesting or is it just shit? And this is where we are at the moment with all of this.'

He shows me into his office at the Gurdon Institute in Cambridge and hefts his tall frame into an office chair, swinging his feet up on the desk. There's no need for formalities here – Eric and I go way back to our time as graduate students together at the Gurdon. Now he is the Herchel Smith Professor of Genetics at Cambridge University and, as of today, has just discovered he's been listed in *Who's Who*. Even this news fails to lift his mood. He's feeling old and grumpy, and clearly having one of those days when you wonder whether a life spent slaving at the coalface of science is really worth it. 'I'm writing a book about how genes work,' I tell him excitedly. 'And I thought you'd be good to talk to.' 'Sounds great,' he replies, in his deadpan German accent. 'When you find out, let me know.' Come on, Eric, I think, at least you could try and sound enthusiastic. Behind him a large tank full of pet fish gently bubbles away, its lamp casting an eerie glow over our conversation as the December sunlight fades from the window of the unlit office. He huffs another sigh.

Eric's depression is understandable. Thanks to the plummeting cost and soaring throughput of DNA sequencing technology, labs around the world are full of enormous machines studded with blinking lights, churning out endless strings of genetic code. When he and I were students at the end of the 1990s, we were reading DNA in tiny chunks, a few hundred letters at a time. Getting the entire recipe for a single gene took months of painstaking effort, matching up tiny overlapping scraps of information to create the complete text. Today an entire human genome can be read in a few days. But rather than clarifying the contents of our genes and how they work, things have got very complicated indeed.

The problem is that we were not put together by an anally retentive designer with the celestial equivalent of thick-rimmed glasses and a hipster beard. Instead, our genome has been sloppily patched together over eons by evolution, the master bodger. And, for want of a more poetic way of putting it, it's full of rubbish.

🐾

The year 1977 is notable, in my mind at least, for three events in the history of genetics.[*] Firstly, there was the launch of *Star Wars* – a film franchise that has taught us much about key genetic concepts such as the battle between nature and nurture, the risks of sibling interbreeding, and the potential of cloning technology. It also witnessed the release of *Never Mind the Bollocks, Here's the Sex Pistols*, proving another important point about biology: you don't need to be technically perfect, you just need to be good enough to get the point across to make something come alive. And finally, it was the year that saw the birth of the DNA analysis technique that enabled scientists to 'read the book of life' (or

[*]It's also the year I was born. Whether that's a notable event in the history of genetics remains to be seen.

other cheesy metaphor of your choice) and decode the information within our genes.

In a neat parallel with the war being fought at that time in the world of consumer electronics between Betamax and VHS video, two different DNA reading methods battled it out for dominance in the early 1980s. In the end, the simpler technique pioneered by British scientist Fred Sanger won out over the alternative protocol developed by US researchers. Fred's method revolutionised genetics over the following three decades, and has only recently been super-seded by modern 'next-generation' technology – the super-speedy high-definition Blu-ray disc to replace Fred's clunky VHS cassette.

Most of us will be familiar with the vision of DNA as a double helix – the twisted ladder structure figured out by James Watson and Francis Crick in 1953, with more than a little help from Rosalind Franklin and Maurice Wilkins. The sides of the ladder are long chains of sugary molecules, firmly attached to each other. Slung in between, the rungs are made up of pairs of four chemical 'letters' called bases, usually known by their initials A (for adenine), G (guanine), C (cytosine) and T (thymine). These are the letters of life. The order that they come in the ladder is known as a DNA sequence, and reading that order is referred to as sequencing. Although these bases are paired up (hence base pairs) – A always with T, G always with C – scientists usually refer to the order of letters on just one side of the ladder as the sequence. The opposite side is known as the complementary strand, a bit like the negative of a photo.

Just as a child learning to read starts with *Spot the Dog* rather than *War and Peace*, science's efforts to decipher DNA started small. As they were developing their sequencing tech-niques throughout the late 1970s, researchers began by reading the genetic makeup of strange viruses that attack bacteria, known as phages. The first whole genome to be published – a heroic feat back in 1976 – belonged to a phage called MS2. It was just 3,500 letters long and contained only four genes, but it was a start. Run, Spot, run!

After a few tweaks to their method, Fred and his team hit upon the technique that would become known as Sanger sequencing, publishing a paper describing their method in February 1977. The technique works by making copies of the particular fragment of DNA you're interested in (effectively a template), each ending at a different DNA letter along the sequence. Importantly, every one of these end letters is labelled with a speck of radioactivity, meaning that it will show up on X-ray film. By separating out all these copies by length and reading off the last letter of each one from an X-ray picture of it, you can infer the sequence of letters in the original template. So if you get 11 letters ending in A, then 12 letters ending in G, then 13 letters ending in T, then 14 letters ending in A, you can assume that part of your original template reads 'AGTA', and so on.

Although it was revolutionary, Sanger sequencing was also painstakingly slow, producing about 100 DNA letters at a time. Over time it became faster, safer and cheaper. Researchers worked out how to use fluorescent dyes instead of radioactivity to label the end letters, developed more efficient techniques for separating the different length DNA strands, and found ways of automating the whole process. Gradually, we got better at reading.

The first lucky organism to have its whole genome read was the bacterium *Haemophilus influenzae*. That sequence was published in 1995, clocking in at just 1.8 million letters compared to our three billion. Yeast followed in 1996 and a tiny worm known as *C. elegans* in 1998. By then, the wheels were well and truly in motion to read our own genetic book. Its chapters are 23 pairs of chromosomes, each a string of DNA ranging in length from about 250 million DNA letters (chromosome 1) to around 50 million in chromosome 22. There are also the X and Y chromosomes that determine your sex – two Xs in every female, but an X and a Y in each male.

By 2001, after a decade of work, billions of dollars and an unseemly tussle between the public- and charity-funded UK-led Human Genome Consortium and brash American Craig Venter's commercial effort, Celera, a rough draft of the human genome was assembled. UK Prime Minister Tony Blair and US President Bill Clinton linked up by satellite to reveal humanity's inner secrets to the world. Much to the frustration of the broadly secular scientific world, Clinton claimed that 'Today, we are learning the language in which God created life.' And lo, there was much hyperbole.

Shortly afterwards, Mike Dexter – then director of the Wellcome Trust, which stumped up a significant proportion of the cash for the sequencing – said, 'When the completion of the first draft of the human genome sequence was announced ... I likened the achievement to the invention of the wheel. I think in retrospect I may have been wrong: I can, just about, imagine technology making the wheel obsolete. Not so the human genome sequence: that will retain its importance as long as humankind exists.'

There's no doubt that putting together a draft human genome – and it was definitely a draft, full of errors and gaps – was a hugely impressive technical and collaborative achievement, and I don't want to make you think that it wasn't. In just five decades we had gone from figuring out that DNA was made of two strands of linked letters, coiled round each other in that famous twisted helix, to being able to read the precise recipes for the multitude of molecules that make us live, breathe, move, think and grow. It took tens of thousands of years for humanity to get to this point, so it was certainly worth a pat on the back and one hell of a party. But when people started to look closely at the contents of our genome in the years following the glitzy announcement, it turned out that maybe things weren't quite what many people had expected. For a start, where were all the genes?

❧

In the year 2000, two men got drunk in a bar at a scientific conference. One was Ewan Birney, who worked at the European Bioinformatics Institute. The other was Francis Collins, director of the US National Human Genome Research Institute. Both were heavily involved in the effort to read the human genome sequence, and together they hatched a plan to run a sweepstake on how many genes humans would turn out to have. It would cost a dollar to bet in 2000, five in 2001 and twenty in 2002, reflecting the increasing analysis that would be taking place, with the winner announced in three years' time at the same conference, held at Cold Spring Harbor in New York. They laid out some definitions of what would count as a gene and started collecting bets, with the conference director jotting each one down by hand in a notebook. By the time of the 2003 meeting, some 460 people had put down hard cash for the GeneSweep. Guesses ranged from more than 150,000 down to 25,000, averaging out at around 60,000 genes needed to make a human. A lot of the higher numbers came in early, based on initial hints that we might have around 100,000 genes.

In the end they were all wrong. Under the rules they'd set down, Ewan and Francis had to pick a winner at the 2003 Cold Spring Harbor meeting. They awarded the jackpot to Lee Rowen, a Seattle genome researcher who plumped for 25,947. The actual number of genes in the database at the time (agreed by the participants in the bet as being the 'correct' figure) was even lower, at 24,847.

Around the time of the first announcement of the draft human genome, David Bentley, head of genetics at the Wellcome Trust's Sanger Institute, confidently announced, 'We will have a best reference sequence by 2003, and will have identified all the genes by 2004. That will give us access to all the genetic information about ourselves – it's unchanging, and unequivocal.' Ah, David. It's tough to make predictions, especially about the future. Most of the researchers I've spoken to recently put the number of human genes at somewhere around 20,000 or so – certainly less than 2 per cent of the whole genome – but this figure seems to be falling. A 2014 study suggested that we might have only 19,000 genes, depending on how they're counted. Whatever the precise number, it's still a lot fewer than we expected, and a mere fraction of the 100,000-ish that was bandied around in the early 1990s when the human genome project first started.

Some of this was to do with collective ego, based on sequencing the genomes of simpler organisms such as the nematode worm *C. elegans* (roughly 20,000 genes) and the fruit fly *Drosophila* (15,000). There was perhaps an assumption that because humans were clearly the most awesome animals on the whole damn planet, we would need a huge number of genes. Not so. Many organisms have far more genes than we do. Water fleas the size of a grain of rice have 30,000 genes – the most of any animal known so far. Plants are particularly blessed in the gene department: grapes have around 30,000, Golden Delicious apples clock in at 57,000, and wheat has nearly 100,000. We're nothing special, at least in terms of the number of genes in our DNA.

There's also the question of what is a gene, anyway? The GeneSweep bookies decided they would only count protein-coding genes in their bet – those containing a precisely defined recipe telling the cell to assemble a particular string of biological building blocks (amino acids) to make a protein molecule. They specifically excluded stretches of DNA that are read into RNA (an intermediary messenger between DNA and protein), but don't actually direct the construction of proteins. As we'll see later on, there are a huge number of these so-called non-coding RNAs produced across the genome, yet little clarity about what they all do and which – if any – of them should fall under the banner of 'genes'.

It wasn't just the lack of genes that was disappointing. To use a TV metaphor, you might imagine that the human genome would be like the glorious best of the BBC, packed full of fascinating and unique programming: some news, late-night films, a bit of sport, documentaries and dramas galore. But in fact it's akin to the worst cable channel, crammed with endless repeats of long-cancelled shows, with only an occasional original to lighten the tedium.

One of the most intriguing results from the human genome project was the confirmation of earlier, less sophisticated analysis suggesting that about half of our DNA is stuffed full of short repeated sequences. As an example, your genome is peppered with about 1.1 million copies of a small repeated

DNA phrase called an Alu element, around 300 DNA letters long, which we share with other primates like chimps and gorillas. Alu, like many of the other repeated elements in the human and other genomes, has come from a transposon – a kind of genetic 'virus' that can randomly copy and paste itself around the genome, making more and more versions of itself. Exactly when and where we picked it up isn't entirely clear, but it must have invaded since primates split off from other mammals around 65 million years ago.

Most of the transposons in our genome are dead. Either they're incomplete or have some kind of mistake, which means that they can't get up and hop around in the genome any more. Scientists tend to view them as a kind of boring stuffing packed around our precious genes, but some elements have been co-opted into forming part of a protein-coding gene, or become part of the control switches that turn genes on and off. Furthermore, as we'll discover in Chapter 16, a handful of them are still very much alive.

So what are we to make of all this stuff in between our genes? Is it just junk?

Many people believe the term junk DNA is the work of Korean geneticist Susumu Ohno.[*] In 1972 he published a paper in the obscure journal *Brookhaven Symposia in Biology* entitled 'So much "junk" DNA in our genome', musing on the following mathematical problem. By the 1970s, scientists had already figured out how much DNA was in a single *E. coli* bacterium and that it contained a few thousand genes. They also knew how much DNA was in each one of our own human cells – 750 times that of the bugs. If the number of genes in a genome was directly proportional to the amount of DNA, then a simple back-of-the-envelope calculation says that we should have around three million genes. Ohno refused to believe it, pointing out that the amount of DNA in the cells of 'lowly lungfish and salamanders can be 36 times greater than our own'. And what would a

---

[*]In fact, there's evidence that the term 'junk DNA' was common parlance in genetics circles back in the 1960s, including by Francis Crick, out of Watson and Crick.

lungfish need with a hundred million genes, eh? To Ohno, all this extra DNA was the result of evolutionary processes at work as parts of the genome get copied and tested by natural selection. If something was beneficial, it would stick around and become a proper gene. If not, well, it stuck around anyway, acting as padding for the genes that did the real work. As he put it, 'The triumphs as well as failures of nature's past experiments appear to be contained in our genome.'

This idea of 'junk' DNA was instantly appealing to journalists, despite protestations from the scientific community that some of it was actually very useful, thank you very much. Yet the name stuck, and it still pops up with alarming regularity in the media. The last time junk DNA hit the headlines in a big way was in 2012, when a large international collaboration of scientists published the results of a massive project called ENCODE – the Encyclopedia of DNA Elements. Co-ordinated by GeneSweep bookie Ewan Birney, this was an audacious attempt to rifle through the non-coding bits of our genome, and figure out what (if anything) they actually do.

According to the ENCODE team, who published their data simultaneously in an impressive 30 research papers, along with about as much hype as it's possible to gather in the world of science, the answer was loads. Loads and loads. Of the non-coding DNA they looked at (which wasn't the entire genome), they found that most of it appeared to be doing something. A lot of it had some proteins stuck to it – the kind of proteins that can switch genes on or off. Some of it was being read into RNA. In total, they concluded that about 80 per cent of our genome was functional in some way. About nine per cent clearly contained control switches for turning genes on and off, while the rest was ... well, it didn't matter what it was doing, but it definitely did *something*.

This was big news – finally we could explain what all this genomic 'dark matter' is actually for. A comment piece in the journal *Science* stated that ENCODE had 'written a eulogy for junk DNA'. Predictably, the science blogosphere went nuts for it, with prominent writers breathlessly repeating the claims. On the less scientific side, creationists and proponents of intelligent design – both flavours of the 'God–did–it' approach to understanding nature – were keen to jump on the findings as proof that our genomes aren't just a collection of pointless junk accumulated over billions of years of evolution. It all does something! We are special after all!

The excitement even spilled over into the mainstream press, with the *Daily Mail* claiming that 'It is a discovery which scientists say could revolutionise the search for new medicines. Vast swathes of our DNA previously thought of as 'junk' could actually hold hundreds of clues to the treatment of diseases.' *The Guardian* was only slightly more circumspect, saying that 'Long stretches of DNA previously dismissed as 'junk' are in fact crucial to the way our genome works.' Then, as with all overhyped celebrities, came the backlash.

CHAPTER TWO

Taking Out the Garbage

Firing up my temperamental Skype connection at the appointed time, I wait for the skittish pixels to solidify into a picture. Eventually I can make out the stern-looking face of a bald-headed older man peering at me quizzically through specs slung on a beaded chain. I half expect him to tap at the screen and ask if this thing is on. Speaking with an unplaceable middle-European accent – 'English is not my mother tongue. I don't have one, by the way – I speak with very bad accents in three languages!' – he holds forth. This is Dan Graur, Professor of Molecular Evolutionary Bioinformatics at the University of Houston, and he has a lot to say about the contents of our genome and its usefulness (or not).

I approach our call with some trepidation. Dan's blog, *Judge Starling*, is a bevy of broadsides aimed at those he refers to as 'glorified technicians and computer jocks' – the hundreds, if not thousands, of researchers involved in large-scale genome analysis projects around the world. He also hasn't been shy about slamming what he considers to be sloppy journalism about genomics, junk DNA and the mysteries within our genome. ENCODE, along with the breathless reporting about it in the scientific and popular press, has come in for a particular pasting.

'First of all,' he says, 'the genome is a product of the natural process – a product of evolution. And therefore it is not perfect or – how shall I put it? – flawless. Moreover, human populations are very small. The fact that there are now seven billion people in the world means very little.' What Dan means is that humans have what they call a small effective population size, which is the number of individuals needed to breed with each other to maintain the necessary amount of genetic diversity amongst themselves. As he patiently explains to me, the explosion in the human population is extremely recent, in the grand scheme of things. Not only that, but we're lucky to be here at all. There's good evidence that the human population went through one or more major collapses in our past, bringing us close to the brink of extinction. One suspect is the eruption of the Toba super-volcano in Indonesia around 70,000 years ago, triggering the equivalent of a nuclear winter and decimating our ancestors (although the latest research casts doubt on this theory). The genetic bottleneck, and others like it in our precarious history, meant that there simply weren't that many people left to make babies with each other. As a result, these squeezes have left an indelible imprint on our genes.

Whether the volcano theory holds up or not, researchers analysing the genomes of modern humans think that our human-like ancestors maintained a breeding population of around 18,500 individuals about a million years ago, making us an endangered species on a par with today's chimps and gorillas. Subsequent changes in our DNA that have persisted and spread through the population are relatively rare, whether that's single letter 'typos' or more substantial chunks of alteration, and we haven't managed to accumulate significant amounts of genetic variation between different groups of people across the globe. Despite the glorious diversity of human beings on the planet, we're all boringly similar at a genetic level. In fact we have about half the amount of genetic variation enjoyed by chimpanzees, while the genomes of tiny nematode worms are 150 times more diverse than our own. This boringness – along with our predilection for near-extinction – means that there is relatively little fodder for evolution to do its work, selecting the genetic good stuff that's useful (or merely putting up with stuff that's not immediately harmful) for our species and getting rid of the crap that isn't.

'Evolution is a very slow process. It's neither efficient nor instantaneous, so bad things, or things that don't matter, can accumulate,' Dan says. 'Since I live now in the United States, the first metaphor that comes to mind is the garage. It's full of junk yet is functional. Only places that sell cars have these very nice neat garages where everything is in the right place.' If you're an untidy person like me, it's hard to admit that the accumulated heaps of stuff piled up on the desk, proliferating in kitchen drawers or lurking in the garage are just useless junk, rather than carefully selected treasures. Similarly, some people have found it hard to believe that our genome is not a show-home for neatly organised DNA, with a place for everything and everything in its place. Yet it's hard to argue that all the junk we have is absolutely necessary when you consider that fugu pufferfish – the infamously poisonous Japanese delicacy – essentially have all the same genes that we do but get by with an eighth as much DNA. Their genome is astonishingly compact compared to most animals, but nobody quite knows why.

The fact that we humans take a long time to breed compared with many other organisms helps to explain why we might be carrying so much genetic junk in our genomes. Fruit flies, for example, are hatched, matched and dispatched in a matter of weeks. This fast turnover can be likened to unfortunate renters having to move house every few months: there's a strong incentive to get rid of as much unnecessary trash as possible. But with our small effective population size and slow breeding, our genomes are more like long-term family homes, crammed with decades' worth of accumu- lated junk.

In the manner of a merely untidy person gazing smugly at a reality TV show about compulsive hoarders, we are by no means the craziest collectors of the biological world. For example, an onion has five times more DNA than you. And that's not the worst offender, according to Dan. 'Currently the biggest known genome in existence is a stupid flower called *Paris japonica* – the Japanese call it the canopy flower. It's got about hundred times more DNA than humans or something like that.' 'But humans want to feel that they're special,' I whine. 'Am I not more special than an onion?' I'm treated to a dismissive eye-roll. 'Either you have to assume that humans are the pinnacle of creation – that everything is functional and those organisms with more DNA than us have junk DNA but we don't. Or you have to assume that humans are a regular organism that has junk DNA just like everything else.'

He brings up the example of Johannes Kepler, the German astronomer who helped convince the world that we revolve around the sun rather than the other way round. 'Humans object every time to their demotion from the centre of the universe. Let's face it, we are not special. People say, "Oh, the dinosaurs went extinct and we are alive!" Actually the dinosaurs ruled the Earth for millions of years and the human race has been here for about 100,000 years, so let's be modest, hmm? Evolution doesn't care about the fact that we write books and get university degrees.'

It's a humbling thought, but he's right. To put it bluntly, evolution only really cares that you get laid and how many babies you make in the process. 'That's the only thing,' he agrees. 'If you look at the world with more objective eyes, you'll see that we are nothing. There are more ants than humans, and in terms of number of species then it's of course beetles.' He tips a nod to the quote, or variations of it, attributed to the brilliant geneticist J. B. S. Haldane, who said, 'If one could conclude as to the nature of the Creator from a study of creation, it would appear that God has an inordinate fondness for beetles.'

Beetles and onions aside, evolution has still decided to put at least some of the junk in our genome to good use. There are an increasing number of examples – some of which we'll look at in more detail later – where DNA sequences that might previously have been dismissed as useless have been put to work. This includes the repetitive remains of long-dead virus-like transposons, or broken copies of existing genes that have degraded over time. These examples make me think of the house-share I used to live in a few years ago. My house-mates were three guys, none of whom were any tidier than me. Letters and small packages would pile up on a rickety shelf over the radiator by the front door while cardboard boxes, emptied of their more exciting cargo, lurked in the hallway. Eventually, after one unopened junk mailing too many, the shelf worked free of its fixings, slumping to an alarming angle and dumping the whole lot onto the wood-wormed floorboards. Rather than going out and buying the proper tools to fix it, we just left it like that. For months.

I was the first to crack. Rummaging in one of the discarded boxes, I found a cube of expanded polystyrene packing material. It was exactly the right size to wedge between the shelf and the radiator, keeping it propped up at the correct angle and restoring its function. It was never designed to be a substitute bracket and should have been thrown away ages ago, but it was in the right place at the right time and it worked. And, as far as I know, it's still there. But just as most of the stuff in the hallway was really rubbish and we were just too lazy to put it in the bin, Dan is unconvinced that most of the junk in our genome is eagerly waiting for its moment of usefulness.

To steal an analogy he used recently on his blog, the claim that any piece of 'junk' DNA could become functional one day is as helpful as saying that any child born in the US could become president. Technically it's true, and certainly a lucky handful of them will, but it's so unlikely for any individual child as to be meaningless. There are probably some situations that make either scenario more likely. Having the surname Bush probably helps your chances of becoming president, while a bit of junk sitting near a functional gene has a better chance of getting co-opted into its regulatory mechanisms. Even so, these situations are extremely rare.

'If you think about something like your house, how many things could you throw out without noticing? Probably a lot.' I glance guiltily around my office stacked with books, papers and large plastic storage boxes filled with I'm-sure-this-will-come-in-useful-one-day stuff. Then I discreetly adjust the webcam so he can't see it. 'I guess the key thing is, what do you define as useful?' I ask, somewhat defensively. This is an important point. When thinking about the role of different bits of the genome, it depends a lot on what we mean by 'functional'. Is it enough to discover that a particular stretch of DNA has a protein molecule stuck to it, or is being read into RNA (both of which happen across most of the genome, and fall under the banner of biochemistry), or does it need to actually do a useful job in the cell? It's like walking into a busy office and seeing a whole bunch of people there, staring at computer screens. Some of them might be actually working – crunching through a spreadsheet or preparing a document – but others are just keeping the seat warm and checking Facebook until it's time for lunch. If I was their boss, I know which ones I would class as 'functional' when it came to appraisal time. Furthermore, just finding someone in an office building doesn't mean they actually work there. They could be a visitor, or merely wandered in off the street. Finding evidence of an interaction proves nothing.

Dan has another analogy. 'As you know, in life, any object does all kinds of things, but that doesn't mean that's its particular function,' he says. 'When you walk in Houston, you invariably step into chewing gum, but that doesn't mean that the function of your shoes is to bind chewing gum. The people working on ENCODE took every activity, every biochemical activity, and said, "This is a function." But that is complete bullshit.'

The problem is that biology is actually organised chemistry. The nucleus of the cell, where the DNA hangs out, is a crowded place packed with all sorts of proteins involved in turning genes on and off. By sheer chance some of them are going to end up in the wrong place, however briefly, and can be detected using modern super-sensitive techniques. Researchers tend to refer to these as 'stochastic interactions', as it sounds a bit more sciencey than 'there's just all this random stuff going on'.

'My analysis of what happened with ENCODE is that they spent tons of money, in excess of about $300 million, and they wasted the lives of maybe 200 postdocs and other people. They had to say something big, they had to say, "We will cure cancer!" or something. I have recently noticed a big similarity between the promises of genetics and the promises in the '60s that we would have flying machines and things like that. It didn't happen. This may be a side-effect of me being old, but it's all public relations now and you have to say something big.

When there are a million papers published every year you have to do something – that's how I see it.'

In Dan's eyes, this disconnect between finding a whisper of an interaction and actually nailing down a proper function for all the stuff in our genome has come about because the people doing these experiments aren't well versed in evolutionary theory. And it simply doesn't support the idea that a huge chunk of our DNA has to be there for a reason. He picks another parallel, this time from the world of physics. 'If you go to London and you suddenly see people flying in the air, and you don't see it anywhere except London, you will not say, "Oh look, there are exceptions to gravity."' Instead, you have to do the hard work of coming up with a new explanation, as Einstein did when he realised that Newton's seventeenth-century equations didn't really cut it when it came to dealing with the complexities of space-time. Dan thinks the same rules should apply to the genome jockeys and their claims. 'If you want to say that, you will have to change the entire body of population genetics and come up with a different theory. But for some reason they think biology is a soft science, and they say, "OK, let's make exceptions."'

To be fair, the ENCODE team have recanted somewhat and pulled back from their claim that up to 80 per cent of the genome does something useful. But this isn't enough to convince the curmudgeon on my computer screen. He tells me about a recent paper from a team in Oxford, suggesting that a mere eight per cent or so of the genome is functional, based on whether or not it's hung around unchanged in our genome for a long time. This easily covers the couple of per cent that actually codes for proteins – our 20,000-ish 'proper' genes – plus a fair chunk made up of control switches and other regulatory stuff. 'I think at this point, we can look at that eight per cent as a minimum. This minimum will go up, but it will never go up to anything close to even 40 per cent, let alone 80.'

Rather than just being curmudgeonly, Dan is trying to bring some clarity to the debate. Instead of dividing all our DNA into boxes labelled 'useful' and 'junk', he's arguing for a more sophisticated approach. First, he draws a line between 'functional' and 'rubbish' DNA. Functional DNA has to definitely do something useful – such as acting as a control switch, being read to make an RNA message that contains the instructions for a protein, or doing something else that's definitely important. This kind of stuff will be preferentially kept in the genome over evolutionary time. To return to my old house-share, out of all the stuff that piled through the front door, these DNA sequences are the birthday cards, bills and purchases that we actually needed to pay attention to.

Then there's all the 'rubbish' – the pizza delivery leaflets, unwanted catalogues, discarded packaging and so on. In DNA terms, Dan divides this rubbish into two further categories – 'junk' and 'garbage'. Junk DNA is OK – you can live with it, and eventually evolution might find a use for it or get rid of it. This is like the handily shaped polystyrene block I used for fixing the shelf, the menu from the one pizza joint that we actually liked, or the boutique catalogues that occasionally tempted me into impulse-buying chichi fairy lights and ill-fitting shoes. It's not that my housemates and I were deliberately keeping this stuff. Rather, we never got round to throwing it out. It's the same for the genome. Just as it takes a lot of time and effort to finally get round to clearing out the garage, unless a genetic element is actually causing major harm – for example, causing a developing foetus or child to die, or rendering its carrier infertile – it can persist in the genome and get passed on, even if it's doing absolutely nothing at all.

And then there's garbage – the stuff that really needed to go in the bin. The junk mail addressed to long-gone previous tenants, the empty envelopes and packing material. That kind of thing. Garbage DNA is positively bad for us. It's the stuff that can still cause unwanted changes in our genome that affect us in a bad way. Eventually, natural selection might kick it out, but for now it persists. As Nobel-winning biologist Sydney Brenner somewhat sexistly pointed out in 1998, 'Were the extra DNA to become disadvantageous, it would become subject to selection, just as junk that takes up too much space, or is beginning to smell, is instantly converted to garbage by one's wife, that excellent Darwinian instrument.'*

The thing that's a bit hard to get your head round is the fact that these categories are not fixed. Our genome is evolving, as it has done for millions of years, and things change. As we'll see, in the same way that bits of junk or even garbage can actually turn out to be useful in the home, junk DNA can become functional if it's in the right place at the right time and does something that benefits the organism it's in. Functional genes can also get copied and broken over time, effectively becoming useless junk. And pretty much anything in the genome can go rogue and start becoming harmful, moving itself into the category for garbage disposal.

There are other ideas about what the function of all the junk might be. Some researchers think that our abundance of apparently useless DNA is the biological equivalent of bubble wrap, acting as protective packing around our genes and helping to dilute out the cancer-causing impact of damaging agents such as X-rays and other carcinogens. It's a controversial idea, though, as many species have much more compact genomes than we do and aren't riddled with cancer. Moreover, some of the repetitive DNA in our genomes is structurally important. This includes the 'caps' on the ends of each chromosome, known as telomeres, which stop them from unravelling. There are also centromeres, which are important for enabling chromosomes to attach to the molecular machinery involved in cell division. So it may be that some of the other junk is also structural, helping to space genes and their control switches out in the right way. But this is also hard to prove. Using genetic-engineering techniques, it's possible to 'glue' a gene right next to the switch that activates it and it still works, suggesting that the precise spacing isn't that important.

---

*Interesting that Brenner never considered his wife might take the option I eventually chose with my messy housemates and simply move out.

Having spoken to a few people on both sides of the 'Our genome is mostly non-functional rubbish/Oh no it isn't/Oh yes it is!' debate, the issue seems to boil down to cost. Not in monetary terms, but rather in the energy the cell expends in terms of sticking proteins all over the place where they're not really needed, or transcribing huge chunks of the genome into RNA for no reason. Some scientists find it hard to believe that cells would go to all that effort of churning out reams of RNA only to trash it, so it must be doing *something*.

Others feel that transcription is cheap. They believe that all this extra RNA from the non-coding parts of the genome is produced as a by-product of reading the important bits, and it would probably cost more for cells to evolve a way to stop doing it. This unwanted RNA just gets torn up and recycled again. By way of analogy, if you're baking cookies, cutting out shapes from rolled-out dough, there will be scraps left over around the edges. But you can mash them back together and make more cookies. There will be a little bit of waste and a bit more time spent cutting and rolling, but it would take you a lot longer to precisely fit the cutter shapes together to minimise it. For now, nobody has actually sat down and calculated the energy costs of these different views so the fight rages on.

In Dan Graur's opinion, problems arise because many people – particularly those who come to biology from the fields of maths, engineering and computing – struggle with this untidiness. Instead, they prefer the clean-shaven sweep of Occam's razor, hacking through the matted beard of complexity to reveal the chiselled jaw of elegant simplicity. Unfortunately biological systems aren't straightforward or simple. They've been built by millions of years of evolution, cobbled together from all the pieces that happened to be hanging around. Dan describes it as bureaucracy gone mad – 'Why do something simple when you can do it very compli-cated, hmm?' – with many overly complex processes that have arisen in order to cope with other overly complex systems. No process engineer would put up with such circular inefficiency, yet our genome seems happy to tolerate this dystopian situation.

To finish, I ask him what one thing he wants people to understand about what's in our genome and how it works. 'I would say stop looking at the genome as a sort of blueprint that is executed precisely and builds a house. The genome contains data. It also contains tons of junk, and it is changing continuously. There is a ton of variation within your body – every cell is different from every other one because it accumulates mutations. The genes are switched on, they interact with other genes, they interact with the environment. It's a complicated thing and we should be very modest in how we interpret it. But we have to look at the genome as a natural product, and stop looking at it as something divinely made to perfection.'

So, that's your genome as we now know it: a genetic garage packed with junk, garbage, control switches and a few thousand 'proper' protein-making genes dotted about in all the mess. But that doesn't explain how these precious treasures actually work. After all this talk of genes, genomes and junk, we're going to need to do a bit of biochemistry. And to do that, we've got to roll up our sleeves and get messy in the kitchen.

CHAPTER THREE

A Bit of Dogma

Although all metaphors for genetics are flawed at best, a good place to start thinking about how genes work at a molecular level is to imagine the genome as a vast library of cookery books. Within it there are thousands of recipes for pies, cakes, soups and more. There are also instructions for components of other recipes: custard for a trifle, a savoury crust for a quiche, buttercream frosting for cupcakes. However, the librarian in charge of this collection is extremely precious about the books, and refuses to let them out of sight in case they get damaged. And, of course, you can't start cooking in a library. If you want to make something, you have to go to the library and photocopy the recipe, then take the copy back to your kitchen. From there, you can start assembling all the ingredients you need and combine them together in the right order to make a tasty treat.

Unfortunately, in the chaos of the kitchen, the photocopied recipe gradually gets covered in splodges and smears and becomes unreadable. So you have to head back to the library for another copy if you want to make it again. And if you need to make a large amount of something – say, a thousand cupcakes for a bake sale – you team up with a bunch of other bakers, handing out hundreds of copied recipes so they can all help out with production.

This, more or less, is the basic process for turning the information written in our DNA (the recipes) into proteins – the stuff in our cells that makes them work (the pies, the cakes and so on). We have structural proteins that give our cells their shape, molecular generators that convert sugars in our food into energy to keep us alive, enzymes that make hormones and all sorts of other biological signals, as well as the biological machines that read, write and repair DNA itself. There are plenty more, and an average human cell makes around 100,000 different proteins.[*] Some of them are vanishingly scarce – maybe just a handful of molecules of a particular protein in a cell – while others are produced in wild abundance. The amount of any given protein that's made is related to how many copies of its recipe are kicking about in the cell, and how efficiently it can be read by the cooks in the molecular kitchen.

Every cell in your body[†] has a more or less complete set of DNA. This is your genome. It's arranged into 23 pairs of chromosomes – each one a long string of DNA – making 46 individual strings in total. The DNA always stays inside the 'library' – a little bag called the nucleus, plopped in the middle of the cell – but proteins don't get made here, just as cakes don't get made in a real library. If a cell needs to make a particular protein, it has to copy out the recipe in the gene and take it out of the nucleus. A molecular machine called RNA polymerase reads the appropriate stretch of DNA and creates a copy of the letter sequence, made out of a chemical called RNA. This is very similar to DNA, and has the same order of letters as the original gene, but it's just one strand – like one half of a ladder that has been vertically sliced through the middle of its rungs.

If there's a need to make lots of a specific protein – for example, a cell in your pancreas pumping out loads of insulin, the hormone that helps to regulate blood sugar levels – then it will make lots and lots of copies of the RNA message from the insulin gene. If it doesn't need so many, then fewer RNA copies will be made. The main thing to remember is that this messenger RNA is the intermediary that carries the information encoded in the gene into the cell to actually build something useful.

Then a spot of processing goes on, as we'll see in detail later. Some bits get chopped out, some get changed, and a cap

---

[*]This is a significantly larger number than the 20,000-ish protein-coding genes we have in our genome. Hold on to that thought – we'll come back to it later.

[†]Except red blood cells, which lose their DNA as they mature.

gets put on one end and a tail on the other, among other modifications. Once all that is done, the RNA message gets transported out of the nucleus to the cell's 'kitchen' – molecular factories called ribosomes. Here, the instructions encoded in the sequence of letters in the RNA are used to assemble building blocks (amino acids) in a long string, in the correct order to make that particular protein. This process of biological cookery is known as translation.

The genetic code itself is a work of beautiful simplicity. Each three letters (or triplet) of the RNA message (also reflecting the DNA sequence of the gene it was copied from) relates to one particular amino acid out of the 20 that occur in nature. These triplets don't overlap – for example, an RNA message containing the sequence AAGTTCGCA* would be read as AAG-TTC-GCA, rather than AAG-AGT-GTT and so on. The three-letter DNA code is the same in all living things we know of, from viruses and bacteria upwards, and is one of the most ancient and fundamental components of life. It explains why scientists can take a gene out of jellyfish or bacteria and put it into mouse or human cells and it will often still work, despite being separated by millions of years of evolution. Under the bonnet, life is all pretty much the same.

Floating about in every cell are many thousands of strangely shaped RNA fragments called transfer RNAs, or tRNAs for short, which act like an 'adapter' between RNA and proteins. At one end of the tRNA is a docking site for a particular amino acid, while the other end has a special structure known as an anti-codon, which contains the complementary triplet code for that building block. For example, if the messenger RNA reads CGC – the three-letter instruction to add an amino acid called arginine – the anti-codon of the tRNA attached to arginine will read GCG (as complementary letters pair up: C with G, A with T). For

*The letter T in DNA is actually a similarly shaped chemical called U (uracil) in RNA – I've stuck with T for the sake of simplicity but be aware that this isn't strictly correct.

the next step, when an RNA message is translated, the ribosome 'reads' each triplet in turn and pulls in the appropriate tRNA that pairs up with it, along with its neatly docked amino acid cargo. The ribosome plucks the building block off this adapter, glues it onto the growing protein string and then moves on to read the next triplet in the message. Repeat this process tens or even hundreds of times and you've built a protein molecule.

While there are just 20 different amino acids, a quick bit of maths tells you that we can make 64 possible triplet combinations from the four letters in our DNA alphabet. Therefore, some of the amino acids must be specified by more than one three-letter group. For example, seeing AAA or AAG in an RNA message tells a ribosome to add an amino acid called lysine to the growing protein string. The amino acids serine and arginine both have six possible three-letter combinations, while two others (methionine and tryptophan) have just one triplet each. There are also three different 'stop' triplets that signify the end of the message. Interestingly, virtually every protein recipe starts with the same three letters, ATG, which is the code for methionine. Nobody really knows why. It just is. I guess everything's got to start somewhere.

The process I've just described forms the heart of what Francis Crick jokingly described as the 'central dogma' of molecular biology: DNA makes RNA makes protein (or rather, information flows in that direction).* Throughout the 1950s and '60s this was broadly believed to be a one-way street. DNA could direct the production of RNA copies of itself, but these RNA messages didn't write themselves back into new DNA. And the information in proteins certainly couldn't be written back into RNA or DNA. While there's a

---

*He later revealed he didn't really know what the term 'dogma' actually meant. Also, to be extremely pedantic about it, the 'dogma' was mostly based on the idea that proteins couldn't encode for DNA, RNA or other proteins – even Crick himself thought his idea was probably too simplistic and that information might go both ways between DNA and RNA.

bit of arguing around the edges* it's broadly accepted that this rule still holds true for proteins. Cells don't have a way of reading the amino acid sequence of a protein and creating an RNA or DNA molecule with the corresponding letters. But the idea that DNA only makes RNA, not the other way round, changed in 1970 with a discovery by two US scientists – Howard Temin in Wisconsin and David Baltimore at the Massachusetts Institute of Technology (MIT) – which won them each a share of the 1975 Nobel Prize for physiology.

They'd been working with certain viruses that use RNA as their genetic information, not DNA. These viruses hijack the machinery of the cells they infect, causing them to multiply out of control and form tumours. Temin and Baltimore independently showed that these viruses were creating a DNA version of this RNA and inserting it into their host cell's genome. The results were controversial, to say the least: the foundations of molecular biology were built upon the idea that DNA could copy itself into RNA, and not the other way round. But it was true – and rapidly seized upon by molecular biologists as a useful lab tool for making more stable DNA versions of flaky RNA messages – highlighting the fact that laying down dogma in biology is a sure-fire way to eventually being proven wrong.

🐾

That's the molecular nitty-gritty. Now let's zoom out and see what a whole gene looks like. Just as a cookbook might have a specific format for all its recipes, protein-coding genes

---

*Some scientists also argue that prions – the peculiar proteins responsible for mad cow disease (BSE) and its equivalents in humans, sheep and other organisms – can transfer information. Prion proteins with a faulty shape can convert normally shaped prions into the rogue, disease-causing form. And they may also have an influence on the genome, encouraging further production of faulty forms. But others say that this is transmission of structural, not genetic, information (which by definition can only involve RNA and DNA), so it doesn't count.

have broadly similar structures, which is how scientists can spot them in the genome. Nowadays the bulk of this work is done by computer algorithms – a task known as genome annotation – although it still needs human eyes and brains to sort the genetic wheat from the chaff.

The first thing to look for is a DNA sequence called a promoter, around a hundred to a thousand letters long. There are certain characteristic DNA phrases found in promoters, with wonderful names such as the 'Tata box' and 'Cat box', which are a dead giveaway. They recruit proteins called transcription factors, which help to unwind the DNA ladder so it can be read, and various other proteins that are important for switching on gene activity. Together, these molecules create an enticing attraction for RNA polymerase, which comes over and starts copying the gene into RNA.

There are other distinctive features, too. Long runs of triplets encoding amino acids, uninterrupted by a stop signal, are a strong hint that a piece of DNA might be harbouring a gene that makes a functional protein. Due to genetic accidents over time, our genome contains numerous duplicate copies of certain genes. Some are still active and make proteins, while others have picked up faults that change an amino acid triplet into a stop, or create some other kind of problem. So although RNA polymerase might copy out the message, it won't be translated into a proper protein.

The other thing to know is that we all have two copies of every gene. As I mentioned earlier, we have 23 pairs of chromosomes in every cell – 46 long strings of DNA in total – and we get one of each pair from Mum and one from Dad. When egg or sperm cells are made, these pairs are mixed around and split up. One chromosome from each pair goes into the newly made egg or sperm, effectively halving the amount of DNA. Then when Mummy and Daddy love each other very much, the two halves come together to make a whole genome again. Because the two copies of each gene come from different sources, there's no reason why they should be exactly the same. There are subtle changes in the recipes that only make a slight difference to the resulting protein – the equivalent of adding the zest of three lemons instead of two to a drizzle cake, adding to the rich variety of life. But there can also be major typos. Just as switching two letters in the recipe for a zingy raw tomato salad could leave you with an inedible raw potato salad, some changes can destroy the function of a protein. Alternatively, a gene or even a whole chunk of a chromosome might just get accidentally deleted in its entirety, like ripping out several pages from the cookbook.

This is why it's useful to have two copies of each recipe in our genome. If one is damaged or incorrect, the other can act as a backup and compensate for the loss. There are a few situations where this isn't the case – sometimes a faulty gene of the pair makes a protein that is actively harmful, for example – but usually it just means that a particular process inside a cell might not work quite so well. The problems really start if both copies are faulty at the same time, caused either by inheriting two broken versions or by the working copy picking up a mistake. In these cases there's nothing left to pick up the slack. A good example of this effect is cystic fibrosis, a disease that causes a sufferer's lungs and digestive tubes to get clogged with thick mucus. If a child is born with one faulty copy of a gene called CFTR, which normally makes a protein that pumps salts across the surface of cells, they're fine. But if they are unlucky enough to have inherited a broken copy of CFTR from each of their parents then they will have the disease.

So another thing to remember is that even though we usually think about a gene as a singular entity, there are always two copies of most genes in each cell. The exceptions to this rule are the genes on the sex chromosomes, known as X and Y. Female mammals have two X chromosomes, while males have an X and a Y. The Y chromosome is a stubby little thing, with only around 450 genes, compared to 1,800 on the X. Last time I checked I was definitely a lady (at least, biologically speaking), so while I have two copies of every gene on my pair of X chromosomes, my genetically male friends only have one set, plus a separate bunch of genes from their Y.

❖

We've all seen the headlines: *Scientists find new breast cancer gene. Hundreds of new autism genes discovered. Is this the gene for alcoholism?* What the previous few thousand words should have taught you is that genes are not 'for' cancer, heart disease or schizophrenia. There are no 'genes for' intelligence or drug addiction. *Genes are simply the recipes that make things in a cell.* Too many journalists are guilty of this lazy shorthand (occasionally including myself, I confess), and it's misleading and confusing.

Using 'gene for X disease/characteristic' language contributes to overly deterministic thinking about our genes and what they do for us. Your genes don't 'make' fingers or eyeballs. They direct the formation of signalling proteins that tell naive cells that they're going to be bones in a growing limb or cells in the retina at the back of the eye. And once those cells know that they're going to do this, they start churning out the proteins that make the structural components of bone or the light-sensitive pigments that enable you to see. And your genes don't 'make you' clever or gay or thin or sick, or anything else. Added together, all the unique variation in your genome subtly alters the proteins you build, which in turn affects how well your cells, and – by extension – your body, work and interact with the environment around them. The only things your genes 'make' are proteins and RNA, which together build your body.

Take the example of CFTR, which I mentioned earlier. Everybody has this gene, and in most people it functions perfectly normally, pushing salts in and out of cells to keep our snot and other secretions at the right consistency. It's only when both copies are broken that it causes disease. So it doesn't make sense to call it the 'cystic fibrosis gene' in its unbroken, healthy state. In fact, when it come to faulty genes causing disease, this example is really not very typical. Frustratingly for doctors, scientists and – above all – patients, there are relatively few illnesses that can be pegged to a single faulty gene, and those that can tend to be rare. And in these cases, individual sufferers can have variable symptoms and outcomes, even if they have the same gene faults.[*]

---

[*]More on this in Chapter 18.

The genetics underpinning common diseases that plague us, including heart disease, cancer, diabetes, depression and many more, are much harder to pin down. Research with identical twins, who (theoretically, at least) have identical genomes and usually share the same upbringing, shows that many of these conditions, as well as traits like intelligence or addictive tendencies, have a significant genetic component. For example, if one of a pair of twins develops schizophrenia, the other has a fifty-fifty chance of also getting it, compared to around a one in a hundred chance for unrelated people in the general population. So our nature, written into our DNA, is clearly very important. But nurture – the environment we grow up and live in – and probably a certain amount of random bad luck play a big role too.

Yet there's more, and here's where it gets confusing. Large-scale studies comparing the genomes of thousands of people with a particular condition to thousands without have thrown up tens or even hundreds of suspicious DNA variations, depending on the disease. Yet these changes only seem to increase an individual's risk by a small percentage, and there's a massive chunk of genetic nature that we currently can't account for. Furthermore, most of these changes aren't even in actual protein-coding genes, but are presumably in the control switches within our DNA that turn genes on and off at the right time and in the right place.

Sometimes a single faulty gene or control switch is enough to royally screw things up, such as inheriting two broken copies of CFTR. But more often these changes just tinker round the edges. Maybe you have a slight change in a protein that's involved in repairing damaged DNA, making it a tiny bit less efficient. In itself this won't cause cancer, but it increases the chances of mistakes building up in key genes that make the proteins controlling when and how often your cells divide, which leads to the disease. It also stands to reason that encountering things that increase the chances of damaging your DNA – smoking, X-rays or UV light from the sun are a few obvious examples – will make it more likely that things are going to go awry.

Alternatively, maybe you have subtle variations in your genes encoding for proteins involved in energy production that tell your cells to make more of them, or they're slightly more efficient. If so, you're more likely to be slimmer than someone whose genetic variations create energy-production proteins that are sparse or sluggish. But none of these could be described as genes 'for' fatness, and there's also still a big chunk that depends on the environment. You could have a metabolism that burns like a furnace, but if you work in an office with a seemingly bottomless supply of irresistible cake then you're probably going to put on weight. In fact this is a bad example, as most of the DNA variations that have been linked to obesity seem to be in or around genes that are active in the brain, rather than metabolism.

More powerful than any supercomputer, our brain ultimately controls our behaviour. Even slight changes in the proteins that control the biological wiring within it can have an impact, however subtle, on how we live and interact with the world around us. We are not solely products of nature or nurture – we're shaped by both, and they are inextricable from each other. By definition, no living thing can exist outside of its environment: the environment is wherever you are, and whatever happens to you. So let's drop the 'nature versus nurture' crap too. Rant over.

Here's the take-home message: DNA contains genes, genes tell cells to make proteins, and proteins do stuff. However, there's one more thing. Right at the start of this chapter you'd have noticed this phrase sneaking in: 'Every cell in your body has a more or less complete set of DNA – this is your genome.' Yet we have the best part of a hundred different organs – brain, blood, fat, skin, liver, kidneys … the list goes on – and each one of those has several distinct cell types within it. It's sometimes said that we have more than 200 different types of cells, but this is probably an enormous oversimplification. This brings us to the fundamental problem in gene control: how do the right genes get used at the right time in the right cells? And how do these genes get turned on and off in response to changes in the body or the world around us? To find out, we need to take a trip back to Paris in the 1960s.

CHAPTER FOUR

Throwing the Switch

Lennon and McCartney. Thelma and Louise. Laurel and Hardy. Jedward.[*] History is full of legendary partnerships, and science is no exception. But while the best-known pairing here is probably Watson and Crick, there's another duo whose names are familiar to any student of biology: François Jacob and Jacques Monod.

In the late 1950s and early '60s, France was a major player on the world's scientific stage. At the centre of the action was the Pasteur Institute in the middle of Paris, with Jacob and Monod right at its heart. Very different but complementary characters, between them they figured out the basics of how genes get turned on and off, kick-starting the entire field of molecular biology. Just as a novice cook might start with boiling an egg or making toast rather than plunging straight into a complex tower of pastry and cream, Jacob and Monod knew that the molecular complexity of organisms like humans was going to make things difficult. So they started small, with microscopic rod-shaped *E. coli* bacteria, which normally live in the gut but can be easily grown in flasks of nutrient-packed broth in the lab. Each bug has a single, circular chromosome, which we now know is around four million DNA letters long, strung through with a few thousand genes. Back in the 1960s, this molecular recipe book was mostly a mystery, yet they did know a few things about what these bacteria got up to with their genes.

Jacob and Monod's work at the Pasteur grew out of post-war research into new antibiotics. This drive was spearheaded by André Lwoff, who would go on to take a share of the pair's

---

[*]Note for non-UK readers: Irish singing duo most recently known for their brief cameo in *Sharknado 3: Oh Hell No!*

1965 Nobel Prize.[*] Together they made a dream team: Lwoff was an expert on microbes and how they behaved, Monod was skilled at biochemistry – purifying and studying proteins in the lab – and Jacob had a detailed understanding of the world of DNA. What drove them was just one question: what happens inside bacteria as they adapt to changes in their environment? Using two different examples in nature – one based on food, the other on viral infection – they set about finding an answer.

Bacteria get the energy they need in order to live from sugars in the environment around them. Their favourite food is glucose, but at a pinch they can feed on other sugars, including lactose. Lactose is transported into the cell and broken down to produce glucose by three special proteins, produced via a trio of genes. In times when there's plenty of glucose around, these genes are switched off. Life is fundamentally a bit lazy: why expend energy reading all the genes and making the enzymes needed to feed on lactose when it's not strictly necessary? But when times are tough and there's only lactose to be found, the genes all get switched on. But how?

Jacob and Monod already knew that the genes encoding the three proteins involved in lactose breakdown were strung together in a row in the E. coli genome. And they also found that when they gave their bugs lactose, all three genes got switched on at the same time. The really clever bit was realising that there must be some kind of molecular 'off switch' that kept all the lactose genes silent when they weren't being used, but could be easily flipped when they were needed. Using a whole series of bacteria with different genetic mutations, the French pair figured out how the switch must work.

Although the precise molecular details weren't known at the time, it turns out that close to the trio of lactose genes is

---

[*]According to François Cuzin, Jacob's first graduate student, whenever anyone asked Lwoff, 'What did you do for science?' he always replied, 'I hired François Jacob!' He told me that he did hear it from Jacob, though.

another smaller gene, which is normally active all the time if the cells are growing in glucose. This gene makes a protein called a repressor – the predicted molecular 'off switch' – that sits on the DNA right next to the three lactose genes and stops them from being read by RNA polymerase. So if the cells are happily growing in glucose, then they don't waste any energy making lactose-feeding proteins they don't need. Effectively, the switch is set to OFF.

What happens if you add lactose to the mix? Nothing – because it turns out that glucose stops lactose from getting into the bacterial cells. The repressor keeps on repressing, and the lactose genes stay off. But if there's only lactose and no glucose, then things change. Without glucose there to keep it out, some lactose sneaks in. It sticks to the repressor, hauling it off the DNA next to the lactose-eating genes. And without the repressor there, RNA polymerase can go right ahead and read them, making the enzymes the cell needs to feed on this new supply of lactose. The switch has been flipped to ON.

Jacob and Monod found a similar situation in operation when they looked at bacteria infected with a simple virus called a bacteriophage. These viruses hijack the cell's molecular machinery to make millions of copies of themselves, ultimately exploding out of their host and killing it like a miniature re-enactment of *that* scene from the film *Alien* (without the teeth). However, under certain circumstances, the virus can smuggle its genes into the bug's DNA and lie low. The shock that awakens it is a burst of ultraviolet light, leading to new virus production and inevitable bacterial bursting. As with the flip between feasting on glucose or lactose, Jacob and Monod wanted to know the molecular nature of this switch between quiet dormancy and messy explosion. Using nothing more than fierce logic and genetically mutated strains of bacteria and viruses, the pair worked out that ultraviolet light beams must be destroying a repressor protein that normally keeps the virus genes in their dormant state. With the repressor gone, the virus switches on a cascade of gene activation – each step making something required for the next – culminating in the assembly of new viruses that burst out of their host.

The discovery that the lactose-eating and virus genes work in the same way – with a repressor being destroyed to turn them on – was startling. To begin with, Jacob and Monod published their findings in the weekly journal of the French Academy of Sciences (*en français, naturellement*). But when it became clear that they had unlocked one of the major secrets of life, they sent a paper outlining their discoveries to the *Journal of Molecular Biology*, for the English-speaking world to read. It came out in 1961, and it was groundbreaking.

Before this point, researchers had believed that genes just made structural things inside cells: proteins that held the cell together or made energy – that sort of thing. This was the first time that anyone had figured out that there was a whole new class of genes making proteins that controlled how genes got turned on and off. The other important thing was realising that genes (at least in bacteria) were self-regulating, controlled by carefully balanced genetic circuits responding to signals from the outside world, such as ultraviolet light or changing food availability. This idea fitted nicely with the rapid rise of electronic engineering, which was also happening around that time. The 1965 Nobel presentation speech announcing Jacob, Monod and Lwoff's shared prize even includes the line 'One can speak here of chemical control circuits similar in many ways to electrical circuits, for example in a television set.' Yet, as we'll see, this 'genetics-as-electronics' metaphor hasn't been entirely helpful when it comes to explaining some of the complexities of biology in real life.

While bacterial gene control turned out to be a relatively simple matter of on/off switches, feedback loops and a handful of regulating proteins – akin to the kind of electronic circuit a reasonably competent teenager could knock together in a physics class – the situation is much more complex in anything bigger. Bacteria are single cells, all more or less identical to each other while living independent lives. But they all need to respond in the same way to the cues and changes around them, such as the availability of different food sources.

As soon as cells gather together to form an organism there are bigger challenges to be overcome. Individual cells have an identity – are you skin or gut, muscle or nerve cell? – and a location – inside or outside, top or bottom? They need to communicate with each other, making decisions about what to do and what to become and then sticking to the plan. In order to do this, the right genes need to be turned on at the right time and in the right cells, without interfering with whatever's going on in their neighbours.

In the intervening years since Jacob and Monod figured out the neat circuits in *E. coli*, scientists have started to unpick the mysteries of gene activation in more complex cells. And it turns out that it's not so much the same thing plus a few bells and whistles. It's the same thing plus bells, whistles, maracas, bongos, vuvuzelas, glowsticks and a dancing monkey with a pair of miniature cymbals. The first thing to realise is this: while gene regulation in bacteria tends to be all about repression – turning off genes when they're not needed – it's more or less the opposite in more complex organisms. A bug's DNA exists as a loosely packed circle in the middle of the bacterial cell, instead of being packaged away into a separate structure in the cell (the nucleus) as it is in more complex organisms. Because of this characteristic, scientists refer to bacteria as prokaryotes, from a Greek word meaning 'before the nut' (*i.e.* 'before the nucleus'). This means that it's easy for the gene-reading machinery to access the DNA at any time, whether a gene is meant to be active or not, so repressor proteins are needed to keep genes quiet when they're not required.

More complicated cells, from yeast upwards, are known as eukaryotes (which translates roughly as 'truly nutty'). They keep their DNA tightly packaged in the nut-like safety of the nucleus using proteins called histones. It's this tight packaging that poses problems for turning genes on: RNA polymerase and its associated transcription factors can't easily fight their way through the packaging proteins to get at the DNA and read the gene. By default, this tends to shut down most genes if they're not needed. But when they are, other proteins have to pave the way for the polymerase, shuffling aside the packaging molecules so the gene can be switched on. In eukaryotes, the challenge is more about turning things on rather than off – it's a matter of activation rather than repression.

Figuring out the molecular ins and out of gene activation is at the very heart of understanding how our genes work, and it's worth spending a bit of time delving into it. As a brand new undergraduate, some of the first lectures I sat through were all about bacterial gene regulation, before we moved on to more complex systems. Our Bible for the course was Mark Ptashne's *A Genetic Switch* – a slim volume of a hundred or so pages, packed with diagrams showing blobby proteins attached to spiralling DNA, explaining how researchers like Jacob and Monod worked out how *E. coli* turns just a few of its thousands of genes on and off. With the French pair no longer with us to share their thoughts about what's going on, I turned instead to the man who wrote the definitive book about their work.

CHAPTER FIVE

The Secret's in the Blend

When I arrive at Mark Ptashne's swanky Fifth Avenue apartment block in Manhattan, right next to Central Park, the doorman is apologetic. The entry phone doesn't work properly, so he asks if I can use my phone to call the apartment. I point out that I'm using a UK mobile and it will cost me five dollars. As we work out the best way to contact my interviewee up on the 12th floor, the man tells me, 'I knew as soon as you walked in the door you would be for him.' 'Is it because I look a bit nerdy?' I ask, laughing nervously. I'm not at my best after a humid September morning spent getting lost in the park – hair frizzed up, glasses slipping down my sweaty nose and carrying my trademark geek's backpack. 'No, no – it's because he always has nice visitors! Tell you what, just head on up and knock.'

Judging by his published polemics against sloppy scientific thinking – and the wryly raised eyebrows of other researchers when I mention I'm coming to see him – Mark has a fearsome reputation. I've spent days on planes and trains preparing for our interview, poring over my well-thumbed copy of *A Genetic Switch* and fighting against the twin fogs of jet lag and an overdose of Californian Cabernet Sauvignon. Luckily, my fears are unfounded. He's charming and affable when I knock on the door, ushering me through with a broad smile. 'Come in, come in! Can I get you a coffee?'

As he vanishes to the kitchen in search of drinks I take a quick look around. In one corner is a full-size grand piano (he later tells me that they had to get it in through the window) topped off with a violin resting on a sheaf of sheet music. The place is littered with golfing and musical paraphernalia, but that's not what catches my eye. Instead, I'm drawn to the large frames hanging on every wall. Now, I don't know much about art, but I *think* that might be a …

'That's a Matisse,' he says, coming through with two steaming cups and following my gaze to the dancing figures framed near the window. 'And that's a Matisse.' He puts the cups down on a large glass coffee table, pointing at the small bronze statue on it. We take a brief walk round the apartment. 'That's a Hoffman, that's Lachaise, this is a Matisse ...' I ask about a delicate oriental fabric wall-hanging. 'Oh, that's just something I liked.'

Art tour over we sit down facing each other across the glassy expanse – he on a voluminous sofa set against a huge window overlooking the park, me perched nervously on the edge of an expensive-feeling chair. I scan around for somewhere to plug in my audio recorder. A neat brown cat is sitting in front of the nearest power socket, looking me up and down before haughtily stalking away. 'Don't mind Zebby ... So tell me why you're here again?' I launch into my usual 'I'm writing a book about how genes work' spiel, curious as to how far he thinks we've come from Jacob and Monod's simple bacterial system. He leans back, arms splayed out across the sofa cushions, and starts to expound.

'So, there are two problems that are closely related. Big problems. One is called development, and one is called evolution. And why are those related? We have to go back to the genes, how they get read to make proteins.' As we've seen, genes aren't 'for' anything except the proteins or RNA they encode, but they have to build all the bits of our body. And relatively small, subtle changes in them are the difference between our layout and that of a chimp or a mouse. 'It turns out that some of the same genes that are required for making a head are also needed for making a foot, or a tail, or whatever. You only have a limited number of genes, but they're used in very complicated combinations, which is why the problem looks so complex. Hey, you notice how scientists always use the word "complex" when they don't understand how something works?' I nod with recognition. It's a standard phrase in almost every single research paper I've ever read about genetics or, indeed, anything else. 'I have a theory that what they really mean is "mysterious", but they can't say that, because it doesn't sound sciencey enough! So the first mysterious thing is the development problem. How do you control the activity of genes so that you get the right combination turned on to make a brain during development, and another combination to make a foot?' He gestures to himself, like he's doing the actions to the kids' song about head, shoulders, knees and toes.

That's a bit of an oversimplification, because your brain or foot is made up of lots of different types of cells. Some genes need to be turned on in all cells. These are known as housekeeping genes, and they do the mundane day-to-day jobs: building the cell's infrastructure, making energy, taking out the trash and so on. But genes producing proteins that give a certain cell type its individual characteristics – whether they're nerve cells or supporting cells in the brain, or skin or muscle or bone cells in the foot – should only come on when and where they're needed as a baby grows in the womb.

'Then there's the evolutionary problem,' Mark tells me. 'We have roughly the same number and essentially the same genes, let's say, as a chimpanzee. And the difference between our actual protein-coding genes and any other mammal's is extremely small. Contrary to what people assumed, a large part of evolution has nothing to do with changing what the genes encode, or the nature of the protein they make. It has to do with changing how they're regulated.' He's warming to his theme, wisps of grey hair starting to lift up slightly at the sides of his head as he becomes more animated. 'So, now we look inside the genome and we see these areas of DNA that encode proteins and we call them genes. Then lying between the genes in the so-called dark matter are these things that we could roughly call the genetic switches. And they determine whether the gene is on or off. But they do so by virtue of what proteins bind to them at any given time.' It's these proteins, the transcription factors, that create an enticing platform to attract the gene-reading RNA polymerase. Once the polymerase is there, it starts copying out the gene into RNA – et voilà (as Jacob and Monod might have put it), the gene is switched on and active. 'The astounding thing is that the basic mechanism by which you turn a gene on is so simple – it's just combinations of proteins binding to these switches. For years and years scientists have been, like "Oh, we always knew that that!", but they didn't.'

It may sound simple enough, but there's a question raised from this idea that genes are turned on at the right time in the right place by a particular combination of proteins sitting on the DNA of their control switches: how do the proteins know where to go? This is relatively easy to answer in a simple way, but a lot more complex – or mysterious – at a deeper level. Transcription factor proteins have a specific shape that enables them to recognise different short stretches of DNA letters, known as binding sites. For example, a molecule called Tata Binding Protein is just the right shape to lock onto any stretch of DNA that reads TATAAA.* This sequence is part of the infamous Tata box I mentioned earlier, and is often found near the beginning of genes. Therefore the transcription factor that binds it is a common part of the combination of proteins that turns genes on.

Other transcription factors are less picky. One called MyoD, which turns on genes that make proteins important in muscles, aims for any sequence that reads CA(any letter)(any letter)TG. Similar proteins closely related to MyoD like the same sequence too, but they switch on genes involved in other processes, from cell division to the human body clock. Individual factors have different preferences – some are more choosy, others are relatively promiscuous and will go with anything as long as it looks vaguely OK. Here's where it gets a bit complicated. The binding sites for transcription factors are usually very short – only a handful of DNA letters long. And, as we've seen with MyoD, they're not always a strict set of letters. Across the three billion letters of our genome, there are bound to be quite a lot of sequences that might suit as potential docking sites. So what brings the right combination of factors to the right gene at the right time?

---

*I always read this in my head like the sound of a fanfare –TA-TAAA.

The secret, to steal a cheesy advertising slogan, is in the blend – the particular line-up of transcription factor binding sites close to a gene. This forms what scientists would call a 'regulatory element', but is basically a control switch. For example, a switch near a gene for making a muscle-specific protein might have, say, two sites for MyoD, two for another related factor and four for something else, all strung together in a particular order in the DNA. This combination of sites might only happen near genes that need to be turned on in muscles, and never occurs close to any genes that are active in different types of cells.

But there's another thing, too. If one MyoD protein has already attached to its favourite DNA sequence, it makes it easier for another MyoD protein to come along and get stuck in, like when someone might only go into a bar if they see their friend waiting inside. Similarly, just as someone who's been stood up on a date would probably decide to go home early, without a friend MyoD quickly loses interest in DNA and wanders off to try its luck somewhere else in the genome. So if there are two MyoD binding sites close to each other in a stretch of DNA, then there's a good chance that a pair of proteins will turn up, especially if there are a lot of them knocking around in the vicinity. But if there's only one, or the sites are too far apart for the MyoD proteins to sit comfortably next to each other, then it's unlikely.

This phenomenon – known as co-operativity – is fundamental to how our genes work. For each gene, there's a particular arrangement of transcription factor binding sites which recruit the correct proteins that attract the gene-reading machinery. As Mark explains to me, 'Co-operativity is probably the single most important idea here, because it explains how you can have only a few elements and you get great complexity. It's got to be the right partners, in the right place. So we might step back one bit and look at the huge overarching principle which applies to a lot of gene regulation, especially in cells like ours. You don't have a different machine to read every gene, for example …' 'Like you just have one set of eyeballs to read every recipe in a cookery book?' I suggest. 'Yeah, but you do need to have some kind of specific signal telling you that you're making the right thing. The thing that nature figured out – it's kind of amazing, actually – is that once you have all the reading machinery, it's just a question of recruiting it to the right place. And to do that we have evolved these very simple little factors that get together and attract the RNA polymerase to the gene.'

This provides a powerful solution to the second problem that Mark posed at the beginning of our conversation: evolution. This kind of set-up makes it plausible that evolution can change species over time – something we'll come back to later – creating the incredible diversity of life we see today. 'You're taking the same genes, but just by moving these sites – the switches – from one gene to another you change when it's active. All the proteins do is bind to the switches and recruit the RNA reading machine to the gene. In higher organisms you can have a whole bunch of these switches. And one will work, and another will work, and so on, and you can mix them up. It's like ... what are those things called? ... Tinkertoys?' One of us is clearly showing our age, as I have no idea what he means. 'Lego bricks?' I ask, hopefully. 'That'll do. So what you do is produce complexity by combinatorial use of these bits. You change the specificity, change the order of the pieces, and you get new kinds of organisms. By the way, did you read my other book? It covers a lot of the stuff we've talked about. I wrote it with Alex Gann – he's a profane, heavy-drinking kinda guy.' 'My favourite kind of scientist!' I admit.

Our time is up. As I get ready to leave he rummages in a cupboard for his latest book, *Genes and Signals*. Like a true fangirl, I fish my own battered copy of *A Genetic Switch* from my backpack and get him to sign them both. I also persuade him to pose with me for a selfie in front of a Matisse before I leave. He does so with polite bemusement. Zebby the cat remains unimpressed.

✦

This idea that combinations of proteins sitting on certain DNA sequences turn on genes raises a fundamental issue in biology, and it's a version of Mark's development problem: if every cell has the same DNA, then cells that will form different tissues must somehow end up with a tailor-made combination of proteins that turns on brain cell- or blood-building genes rather than liver- or skin-related ones, for example. And if there are all the same proteins in all of the cells, then they'll switch on all the same genes at once.

So we find ourselves with a literal chicken and egg situation. In order to only turn on the right genes for its type, a cell needs to have already made the correct combination of transcription factor proteins (*i.e.* turned on the genes that make the transcription factors). But to do that, it needed the right combination of *other* factors to turn on *those* genes. And so on, backwards and backwards and backwards until you get to a single egg cell. So how does *that* work?

The best way to explain what's going on is to use the analogy of a video game. This one is a multi-level adventure quest, and the object is to save the beautiful/handsome princess/prince/gender-neutral person of your choice. Of course, you can't just start with the final daring rescue scenario at the enchanted tower. That's the hardest level, and you haven't collected enough weapons, medical packs or skill at pushing the right combination of buttons to get through it. Instead you start at the beginning. Just you, on your own, in a dungeon. You play the level, collecting a few bits and bobs, working out some simple gameplay – run, jump, open a door – and eventually you've done it. Level up. The next level is a haunted forest. Same thing here: find some more useful stuff – remember to grab the magic scuba mask from the wizard! – and gain new skills like rope-swinging or tree-climbing. Level up. Then it's on to the lake floor, which is why you needed the diving gear, then after that you go up the frozen mountain, all the while picking up the bits you need for future scenarios. Eventually you make it to the tower, braving the rotating knives and whatnot to release the captive that awaits you.

In many ways, this game is like the development of a human being – or any other organism – from a single fertilised egg cell to a fully grown baby with all the correct organs and tissues it needs for life. But this is the most complicated video game ever made. Right from the start, the egg is packed with all sorts of goodies, including transcription factors and RNA polymerase, sitting there on the control switches of genes that are important for the very earliest stages of development, just ... waiting. Then the sperm enters, and it all kicks off. These expectant, poised genes – making proteins that are important for creating stem cells in the early embryo – get switched on, and the fertilised egg splits in two. Then it divides again to make four cells, then eight, then 16, and so on. After a few divisions, there's a small ball of stem cells that are more or less identical. But deep down, they're not all exactly the same. This patchwork effect – known as heterogeneity – is in place right from the start. An egg cell is a complicated bag of unevenly distributed biological stuff, not a perfectly homogeneous mixture, and there might be slight random differences between different bits of it. Down at this kind of scale, even tiny fluctuations matter.

The cells in the very early embryo are a bit like players starting off a video game. They're trying stuff out and seeing what they can do. Some of them may have been lucky enough to get a bonus right at the start – perhaps a few extra molecules of something useful ended up in a cell as it divided, compared to its neighbour. And some of them might chance upon a neat trick – for example, making slightly more of a particular transcription factor. As a result, they can progress to the next 'level' of development – maybe starting a journey towards being the head – turning on the next set of genes to make proteins that send them in that direction – while the cells next to them have to make steps towards becoming the tail. But just as in our video game, completing the first level – deciding to become top rather than bottom – doesn't mean that the 'head end' cells can automatically skip straight through to the final stage of making a brain. There are multiple levels in between. All the while cells are trying things out, communicating with their neighbours about which fate to pick and activating the appropriate set of genes for the next stage, based on the kit and skills they gained in the last one. It's a one-way route through the game, and there are no short cuts – at least, under normal circumstances.

As Mark Ptashne mentioned, not only do the control switches that turn genes on help us to make the right bits in the right place at the right time, they also play an important role in shaping our whole species. They're like Lego bricks that can be swapped around to build wonderful new creations from the same underlying parts. And when they go wrong, the results can be dramatic. With that in mind, it's time to meet Hemingway's cats.

CHAPTER SIX

Cats with Thumbs

Why do cats stare when you're pouring milk? It's because they know it's only a matter of time. Time — the only thing between them and opposable thumbs. Imagine that: cats with thumbs.
Cravendale milk advert, 2011

This book is all Bob Hill's fault. I conceived the idea for it during his talk at the Royal Society's meeting on gene regulation, when he showed the photo of Ernest Hemingway's six-toed pets. Two years later, braving sub-zero winter temperatures and rattling over Edinburgh's cobbled streets in a bone-shaking taxi, I've finally come to see a man about a cat.

'If you go up into the northern part of the US, up the eastern seaboard towards Nova Scotia, you'll find about 10 to 15 per cent of the cats have extra toes,' Bob tells me, pouring hot water onto several teaspoons of instant coffee in the mug in front of me. He leans back in his chair and peers at me kindly through his glasses. I suddenly feel more like a child at story time than a science writer interviewing a professor of genetics. I curl my fingers round the cup, easing its warmth into my cold bones, and listen to his tale. 'In the 1700s, a time of great seafaring, the captains liked to have cats on their ships. They particularly liked the six-toed cats because they were supposed to be better mousers – I saw a YouTube video showing a six-toed cat using his extra toe almost as a thumb, so perhaps they could grab hold of the mice better. The other reason was apparently they didn't slide off the boat as easily. Maybe because they had extra toes they could grab hold of things.' I wrinkle my nose sceptically at the thought of these alleged super-powers. Somehow, the idea of a cat grasping hold of a passing rope before it shoots overboard doesn't seem plausible. 'Yeah,' he admits, 'sailors are really superstitious. I think they were probably just good luck charms.'

The story goes that an old sea captain gave Ernest Hemingway a six-toed cat named Snow White, or possibly Snowball. Her prowess as either mouser or deckhand is unknown, but today a clowder of her descendants – many with extra digits – roams Hemingway's estate on Key West, the last in the chain of islands that dribbles from Florida's southern coast. They're as much a part of the attraction as the old man's house itself.

When you see these big-pawed polydactyl creatures – or see humans with extra fingers and toes, like the baby in the large colour photo pinned to Bob's office wall behind him – the immediate assumption is that it's a fault in a gene. But there is no gene that specifically makes a toe, a finger or a thumb. Instead, it's a mistake in a control switch for a gene called Sonic Hedgehog. And yes, it was named after the video-game character.

It all started back in 1980 when two German scientists, Janni Nüsslein-Volhard and Eric Wieschaus, noticed that fruit-fly maggots with a particular gene fault were unusually stumpy and covered in bristles. In the great tradition of fly geneticists everywhere, they picked a name for the gene to reflect what these unfortunate creatures looked like: hedgehogs. Then in the early 1990s, researchers found three versions of the hedgehog gene in mammals. The first two got relatively sensible names: Desert Hedgehog and Indian Hedgehog. But the third was named by Bob Riddle, one of its discoverers, who spotted Sega's famous blue hedgehog in his daughter's comic book. Despite grumblings from the scientific community that calling a gene Sonic Hedgehog was more than a little immature, the name stuck. There's even a rather sweet story about one of Riddle's colleagues being surprised to see that the new-found gene was so important that McDonald's were basing Happy Meals on it.

The Sonic Hedgehog gene carries the instructions telling cells to make a small protein molecule, also called Sonic

Hedgehog. It's one of a multitude of signalling molecules that seep out of cells, passing on important messages to their neighbours like tiny biological couriers. Although Sonic Hedgehog is active in various places at different times as an embryo grows in the womb – in the limbs, the face, the precursors of teeth, brain cells and more – it always does the same job: helping cells make decisions about what to become. Will they be muscle or skin? Tooth or jaw? Little finger or thumb? In this way, the intricate structures of a baby's body are moulded from unformed blobs of naive tissue.

By the late 1990s, Bob and other scientists suspected that Sonic Hedgehog was important for directing the growth of fingers and toes in the right place at the right time. But they had no idea how it worked, or where the control switches were. Then there was a lucky break. A graduate student called James Sharpe, working at the Medical Research Council's Mill Hill labs in north London, noticed something strange. He was studying Hox genes – a group of related genes that are responsible for setting up patterns in the developing embryo. They ensure that animals have a head at the top, feet at the bottom, and everything else in the right place in between. As part of his investigation, Sharpe was creating genetically engineered mice carrying extra copies of a particular Hox gene, randomly inserted into the genome. As well as the effects he expected to see from the additional gene, he also noticed that some of the mice had oddly shaped feet with extra toes. He called them Sasquatch as a nod to the mythical Bigfoot.

One day, Bob Hill happened to be chatting to Sharpe's boss on the phone. 'He said, "We've got this funny mouse that's got extra digits. Are you interested?" Of course I jumped at the chance. So I said, "Send 'em up to me and we'll have a look."' As luck would have it, Sharpe's extra gene had settled itself down close to Sonic Hedgehog. Not that close, though; it was still about 800,000 DNA letters (800 kilobases) away, and it seemed crazy that something so far away could be having an effect on gene activity. All the control switches that had been found so far were around 50 kilobases away

from their genes – this one wasn't even within shouting
distance, molecularly speaking. Also, it was right in the middle
of another gene that's switched on all the time throughout the
body, not just in the growing paws. But there were no other
genes around that were remotely implicated in limb development, so Sonic Hedgehog had to be involved. The question
was – how?

'The next thing we did was to get hold of the DNA next
to where James Sharpe's extra gene had gone in, as we figured
that might be the control element,' Bob explains, scribbling
intricate little diagrams with a sharp pencil on the pages of
a notebook for me. Bob and his postdoc Laura Lettice stuck
that fragment of DNA to something called a LacZ reporter
gene, which makes an enzyme that can cut up certain types
of sugar molecules. It's a powerful and widely used technique
for hunting down the control switches that turn genes on:
simply attach your suspected control element to the LacZ
gene, creating what's known as a transgene, and then make
genetically modified mice carrying that transgene within
their DNA. The idea is that wherever and whenever the
control switch is normally active in a developing mouse
embryo, it will be tricked into turning on production of
LacZ. Washing the embryos in a special sugar-based chemical, which releases a bright blue dye when it's chopped up,
reveals exactly where this potential control element is exerting
its effects (or not). For example, if it's active in a particular
area in the developing paws, then that region will turn blue.
Laura's results were striking.

'This was really lucky,' Bob tells me, scratching a rough
pencil sketch of a developing mouse paw – a paddle-shaped
wad of skin-like tissue known as a limb bud. He also draws
the gene arrangement as a straight line of DNA with neat
boxes sitting along it. 'We found that this one particular
region' – he taps just to the left of James's accidentally
inserted Hox gene – 'gave this beautiful pattern that looked
just like the normal activity of Sonic Hedgehog in the
developing limb.' Scratch, scratch, scratch. He scribbles
shading across the lower side of the blobby bud. If you look

at your own hand, this roughly corresponds to the squidgy part running from the base of your little finger down to the wrist. Straight away Bob knew they had found the elusive control element responsible for turning Sonic Hedgehog on in the developing limb, even though it was such a long way from the gene itself. And once they'd pinned down the region that housed the switch, the next step was to read the DNA sequence of that part of the mouse genome.

The control switch itself turned out to be just 800 letters long. It's virtually identical to a stretch of DNA in the corresponding region of the human genome, around a million base pairs away from our own Sonic Hedgehog gene. It's also found in chickens. 'You get these six-toed Silkie chickens. They're real nice,' he says, looking for a photo on his computer and failing to find one. 'If you can get such a thing as a pretty chicken, those are pretty chickens. Also horses can have extra toes.' I raise a query at this point – surely horses have single hooves, not multiple toes? 'A horse is actually standing on its middle toe.' He (accidentally, I'm sure) gives me the finger, flips his hand over and presses it into the table, laughing. 'Some horses have an extra toe just hanging off above the hoof. It's not a complete spare hoof, but it's almost like a thumb. We haven't looked at DNA from these horses so we don't know if it's the same mutation as the cats, but I'm sure it will be.'

Weirdly, Bob also found the same control-switch sequence in fish. And they *definitely* don't have fingers. 'We found that we could take that short region from pufferfish and use it to replace the one in a mouse. They would still grow fingers just fine. The fish sequence can do the same thing as the mouse sequence.' This isn't unheard of in biology. Scientists have found other examples of genes and control switches that can be swapped about between organisms and still basically do the same thing. Whatever this control region does, if it's preserved so widely across species – and therefore goes a long way back in evolutionary terms – it must be very old and very important. And this means that mistakes within it are likely to have serious effects.

Bob abandons the notebook and turns back to his computer, pulling up a PowerPoint slide covered with neatly stacked rows of DNA letters, each corresponding to the sequence of the Sonic control switch from a different organism. They all seem to be the same, except for a few highlighted columns. 'Here's the Hemingway cat mutation,' he points out, tapping on the screen with the pencil. In one place, the letter A has been switched to a G. One single change in an 800-letter-long control switch far, far away from a gene gives these cats their thumbs. Bob and his team have analysed DNA from the Hemingway cats and other six-toed moggies up and down the United States, and they all share the same genetic fault. It's particularly common in Pixie Bobs – charming tabbies with bobbed tails – as well as Maine Coons, the giants of the cat world. British cats with thumbs have one of two different DNA mistakes. One affects cats in the south, the other creates multi-toed Midlands mogs, but both lie very close to the Hemingway mutation.

Bob pokes at a few more letters on the screen, this time at the lines of human DNA sequence. 'You see where this T is changed to a C? That's from a Belgian family. Every member of the family with extra digits has that change.' 'So what's going on?' I ask, keen to understand how these tiny changes can cause such dramatic effects. 'I dunno!' he shrugs. 'Well, we do have a little bit of an idea.' Bob scratches again in the notepad with the pencil, explaining how the normal control region turns on Sonic Hedgehog in a small area of the lower part of the limb bud as an embryo grows in the womb. This somehow communicates to the developing hand or paw to make a certain number of fingers in the right order, thumb to pinky. If the control switch is completely missing, then no limbs form at all. He shows me some pictures of mouse embryos with this particular genetic fault, each bearing sad little stumps in place of their front and rear legs. But if there are tiny mistakes – like those seen in the Hemingway cats or human families – then something odd happens. An extra band of Sonic Hedgehog appears along the top of the limb bud, right where the thumb should be growing. This double dose creates confusion in the developing limb and it starts growing extra digits. But it's not just a matter of switching on activity of Sonic Hedgehog in the right place; it's also vital that it goes on – and off again – at the right time.

As we grow in the womb, the bones of our fingers and toes form inside each limb bud. When the time is right, the cells in between the digits die away, leaving us with a perfect set of wriggling toes and fingers. But if Sonic Hedgehog hangs around too long, the in-between cells don't die, leaving the fingers fused together. In humans, mice or other animals with separate fingers and toes this is obviously a bad thing. But there are other animals where having skin in between your fingers is a positive advantage, such as *Fledermäuse* – flying mice, as the Germans refer to them. Or bats, as they're more commonly known in English.

'We thought that bats might be interesting,' Bob tells me. 'They have extremely long fingers and the wings are basically webbing between them.' To find out more, Bob set up a collaboration with Nicola Illing at the University of Cape Town in South Africa. So far she's found differences in the Sonic control region between bats with long wings and those with short, stubby wings. There are also key differences between bats and mice, both in the DNA sequence making up the control switch and in the pattern of Sonic Hedgehog gene activity over time as a baby bat's limbs grow while it's in the womb. Then Bob did the obvious experiment: replacing the Sonic control switch in a mouse with one from a bat. 'We were wondering if we'd end up with mice flying around the lab, so that we'd have to keep the lid on the cage to stop them escaping! That didn't happen,' he says, with a slight tinge of disappointment, 'but the results are interesting and we're still trying to figure out what's going on.'

For now the mystery is unsolved, and our time together is up. It's only as I'm about to leave that I notice Bob's mug is adorned with a cartoon hedgehog and the Edward Monkton poem *'As soon I start thinking that I'm sensible and sane, the random hedgehog comes along and fiddles with my brain.*' Bob tears the scribbled pages out of his notebook and hands them to me. On the train back home to London, hammering my way through several cans of unpleasantly warm gin and tonic and a packet of Percy Pigs, I get them out of my bag and take another look. All I see is a jumble of little boxes arranged on horizontal lines, representing the control switches littering the DNA around the Sonic Hedgehog gene, along with bulging limb buds and indecipherable squiggles. The random hedgehog has had his revenge.

❀

The story of Sonic Hedgehog and the Hemingway cats illustrates a few important principles about the way that our genes work. For a start, things are clearly far more complicated than the simple control elements in bacteria. The Sonic switch is a huge distance away, in molecular terms, from the gene it activates. How on earth does it manage to find the start of the gene to make sure it gets read into RNA messages? The exact details of how this happens are not entirely clear, but scientists probing the three-dimensional organisation of DNA within cells are starting to find some clues, as we'll discover in Chapter 9.

Another thing to notice is that the region Bob and his team found is only responsible for activating Sonic Hedgehog in the developing limb. This single purpose is common to many of the control switches that activate individual genes, and there are other switches scattered around near Sonic that specifically turn it on in other places where it's needed. Each of these elements responds to certain signals at specific times and places. For example, the exact combination of signals coming into cells in the lower part of a mouse embryo's paw leads to a particular combination of proteins that sticks to the limb control switch, turning on Sonic Hedgehog just where it's needed, so it can tell neighbouring cells to become the mouse equivalent of a pinky or ring finger, rather than a thumb. A different set of signals in the growing nervous system will lead to a new combination of factors attaching to the Sonic Hedgehog brain switch, turning it on in just the right place to tell cells to become one type of brain cell rather than another. These switches are effectively acting as the sensors of the cell – gathering the information from the signals that come in and interpreting it by way of turning specific genes on. Yet in many cases, including the Sonic switches, the identities of these different groups of proteins – and how exactly they work to activate the gene – aren't yet known.

There's also an intriguing complication. The Sonic control region is slap bang in the middle of a totally different gene that's turned on in all cells all the time, known as a 'housekeeping' gene. Given what we know about how the control switches in the genome work – by bundles of various proteins attaching to them to attract the attention of the transcription machinery – it's hard to see how the housekeeping gene can keep working properly, but it does. This interloper is also deaf to the highly specific activity of the switch that precisely turns on Sonic Hedgehog in a narrow region of the developing limb – the housekeeper just keeps on doing its thing at the same level in all types of cells, all the time. How this happens is a complete mystery, although it's not unique to Sonic. Other genes have their switches located in similarly strange places, and they still work just fine.

There are plenty of unanswered questions surrounding the Sonic control switch and its role in directing the growth of our fingers and toes. And that's just one region regulating one aspect of one gene – 800 letters of DNA out of three billion. There's a lot of work to be done to figure out how the thousands of other switches work to control the activity of the rest of our 20,000 or so protein-coding genes (not to mention all the non-coding RNA, which we'll come to later).

More broadly, the work of Bob's team and others shows that small changes in genetic control switches can have profound effects. A single DNA letter change – from an A to a G – gives the Hemingway cats their thumbs.* And it's not just about making changes that we might consider to be unusual or unwanted. This ability to create dramatic changes in body shape from tiny genetic alterations has been rich fodder for evolution. Just look at the sticklebacks.

*It's tempting to speculate that they're only a few evolutionary steps away from being able to pick locks, load weapons and kill us as we sleep. I'm more of a dog person.

CHAPTER SEVEN

Fish with Hips

Three-spined sticklebacks are plucky little fish, just a few centimetres long. They normally live in the ocean, but every spring they make the epic journey up winding coastal streams to breed in fresh water. Unfortunately, this migratory lifestyle got them into trouble at the end of the last ice age. As the glaciers retreated from the coasts of North America, Europe and Asia, groups of sticklebacks got trapped in newly formed lakes with no way back to the sea. Since then, they've had about 12,000 years or so to get used to it, with each lake's population of fish evolving independently to adapt to the conditions of their new home. As a result, the various populations of fish living in these lakes have developed an astounding array of physical differences. There are now 30 different species, each distinct in their own way and strikingly different from their sea-dwelling cousins.

These curious fishy ex-pats appealed to David Kingsley at Stanford University, who figured that they might reveal some answers to a question that had been bothering him for a while: how do genetic changes produce the differences that we see between species? Is it changes in the actual coding sequences – the recipes for the proteins that build the body – or is it in the control switches? I heard him speak about his work at the same Royal Society meeting where I first encountered Bob Hill and the Hemingway cats. He's from the plaid shirt and beige trousers school of scientists, with what I think must be an East Coast accent despite his California zip code, and his talk begins with a story about fish hips.

Sea-dwelling sticklebacks have a pelvis-like structure towards the end of their backbone, which attaches to a pair of fearsomely spiky hind fins. By wiggling their hips, the sticklebacks can raise their fins up to defend themselves against predators. But fish from the predator-free environs of a land-locked lake don't have these weapons. And when Kingsley looked closer, he noticed that they also lacked the underlying pelvis.

To go from a stickleback with a pelvis to one with none at all in just a few thousand years is an impressive change. So how did it happen? In search of answers, Kingsley and his team started delving into the sticklebacks' genomes, looking for particular regions of DNA that were always associated with either having or lacking hips. They pinned down one in particular and discovered it looked a lot like a gene called Pitx1, which had been previously discovered in mice, humans and chickens. The 'Pit' part of the name comes from its role in the development of the pituitary gland – a pea-sized blob that sits at the base of your brain and pumps out hormones involved in a whole range of bodily functions, from growth and metabolism to sleep and sex. This seemed like a bit of a random connection until it became clear that Pitx1 was actually an alias of a gene called Backfoot, discovered by another lab around the same time. Backfoot is switched on in the back legs of a mouse embryo as it grows in the womb (as well as in the developing pituitary, under its other guise), but not the front limbs.

It's certainly a closer leap to imagine that a mouse's hind legs and a fish's hips might be somehow related rather than trying to make a connection with the pituitary gland. So could it be the case that the lake-dwelling sticklebacks without a pelvis were lacking the Pitx1 gene? Results from experiments in mice weren't promising. Removing Pitx1 did give the animals small limbs, but they also died at birth with a whole range of other problems. In David Kingsley's view, getting rid of Pitx1 entirely seemed like an overly drastic solution for removing the pelvis from a stickleback. This hunch was confirmed when he took a closer look at the DNA encoding Pitx1 in both types of sticklebacks. The gene was identical in both – perfectly normal and functional in the ocean- and the lake-dwellers alike. Something else must be responsible for the change from hips to no hips.

Next, Kingsley and his team looked more closely at the fish themselves to see exactly where the Pitx1 gene was switched on, using a technique that highlights where specific RNAs are being made in developing embryos. There was one significant difference: the marine sticklebacks had tiny spots of activity in the area of the embryo due to become the pelvis, but these were missing in the freshwater ones. So it wasn't a change in the gene itself that gave the fish hips or not, but a difference in when and where it was switched on. To discover what was responsible for this change, Kingsley zoomed in on the DNA sequence next to the Pitx1 gene itself. He and his team started cutting this up into smaller pieces, attaching each one in turn to a gene originally stolen from jellyfish, one that makes an unusual protein that glows green under ultraviolet light.* Then they injected these constructions into fertilised stickleback eggs, to see whether any of the DNA fragments could switch on the green gene in the same places as Pitx1. One of them worked, giving neat glowing green spots in the fish embryo exactly where the pelvis was going to form. To prove this was definitely the switch responsible for the fishes' hips, they attached that specific fragment of DNA to the Pitx1 gene itself, and injected it into the eggs of lake-dwelling sticklebacks. In an impressive demonstration of the power of these elements, the normally hipless fish grew a pelvis – the first of their species to do so for thousands of years.

As a clincher, when the team looked at the DNA sequence of the hip switch in both types of sticklebacks, they discovered that lake-dwelling fish species that had lost their pelvis were all missing this crucial region, regardless of where in the world they came from. They had all mislaid it in different ways – some were missing bigger chunks, others smaller ones – but every species without hips was lacking part of this crucial region in slightly different but overlapping places. And they had all done it on a timescale of thousands, rather than millions, of years.

*Called, unsurprisingly, Green Fluorescent Protein, or GFP.

Kingsley's sticklebacks tell us two things about how evolution shapes organisms over time. Firstly, that as well as the creepingly slow changes that many people associate with evolutionary processes – the oh-so-gradual morphing of fins into limbs over millennia as our ancestors heaved themselves out of the ocean and onto dry land – a single, small genetic change can have a dramatic impact. One little region of the genome gets accidentally deleted and boom! The entire pelvis is missing. If this kind of genetic accident happened to a fish living in the sea, the chances are it would quickly get eaten up before it got a chance to breed, as its hips and associated spiny fins are a vital part of its defences. But in the predator-free environment of a mountain lake, who needs 'em? Secondly, assuming that the new mutation doesn't prevent the fish from getting lucky – either by physically affecting its ability to reproduce or making it look so weird that no other fish want to have sex with it – then there's nothing to stop it from gradually spreading through the population. And if the new variation brings a survival advantage, it'll spread even faster.

🐾

It's not just fish hips and cat thumbs that are the result of small changes in genetic control switches. David Kingsley has also discovered a few human traits that work in the same way, with the most immediately obvious being skin and hair colour. A few years back, he and his team discovered that the DNA around a gene called Kitlg, found in many animals including both sticklebacks and humans, seemed to be chock-full of control switches. The protein encoded by Kitlg (known as Kit ligand) is a biological multi-tasker, helping to make blood, sperm and cells packed full of the dark pigment melanin. It's this molecule that determines your coloration. More melanin and you'll be darker, less and you'll be lighter. Kingsley and his team discovered that playing with these switches in sticklebacks changed their coloration, making them darker or lighter depending on which ones were missing. So they took the same DNA region from humans and broke it down into pieces, testing each one to find out when and where it was active.

Sure enough, they tracked down one specific control switch that could turn on the gene only in skin and hair. Then when they looked at the DNA sequence of this switch in West Africans and white Europeans, they noticed a consistent difference in a region more than 300,000 letters (300 kilobases) away from Kitlg. Not as far as the distance between Sonic Hedgehog and its limb control switch, but still a long way off. One single letter was switched: an A in the Africans, a G in Europeans. Just one.

Next, they tested whether this change affected how well the switch could turn on Kitlg, by looking at the two different versions in skin cells grown in the lab. They discovered that it wasn't as simple as an on/off (or rather, black/white) switch. Instead, the version in Europeans wasn't quite as effective at activating the gene as the African version was. A quick calculation in their paper suggests that having two copies of Kitlg with the European switch makes a person's skin around six or seven shades lighter than someone with two West African versions. Because you have two copies of every gene – one from Mum and one from Dad – the effects of the switches will be more apparent if they are both the same, while having one of each will give a colour somewhere in the middle.

However, there are around 30 shades between a typical Nigerian's dark skin and a pale European complexion, so the difference in the Kitlg switch only explains part of our skin colour, rather than the whole thing. David suspects that there are probably other similar genes and switches out there that add up to give each person their particular hue. But even so, just a single letter can make a big difference to what you look like. This is true of hair colour as well as skin. In 2014 Kingsley and his team published another paper showing that European blondes have a single letter difference in a control switch around 350,000 DNA letters (350 kilobases) away from the Kitlg gene, compared to dark-haired people. Again, it's a tiny change miles away from the gene, but it has a big impact. This subtle alteration in blondes means that a transcription factor protein called Lef can't stick quite as well to the DNA of the control switch, so it's not as effective at turning on Kitlg activity. It's not on/off, but it's enough to significantly cut down the melanin production in hair cells, and make them fair.

Growing up in the 1980s, I would often hear jokes about blondes being stupid – and as a brunette (to my shame) I would often repeat them. I now know better, but many people apparently don't. In a news article about his hair colour research, Kingsley attacked this long-held stereotype, saying, 'It's clear that this hair colour change is occurring through a regulatory mechanism that operates only in the hair. This isn't something that also affects other traits, like intelligence or personality. The change that causes blonde hair is, literally, only skin deep.' Blonde jokes aside, his work on coloration has more profound implications. As I'm writing this chapter, the United States is fracturing under the stress of racial tension following several high-profile incidents of white police officers killing unarmed black people, and a horrific racist shooting in a church. Countless numbers have been unfairly judged, oppressed or killed throughout history because of the colour of their skin, yet it boils down to little more than a handful of DNA letters in a few genetic switches. For a species named after our intelligence – *Homo sapiens* translates as 'man who knows' – we really are very stupid at times.

Moving away from skin colour, an altered genetic switch is also responsible for a less immediately obvious characteristic, which can be found down the trousers of your nearest man. The males of some animal species, such as chimps and mice, have hard little spikes on the base of their penis, known as spines. Exactly why they're there isn't clear, but some animal sexologists have suggested that they might be useful for the kind of rough and ready, wham-bam-thank-you-ma'am sex favoured by highly promiscuous species. Chimps will use their spiky boners to bang anything that moves, compared to the spineless penises of (relatively) more monogamous human males.

While searching through a catalogue of DNA regions that were missing in humans but present in chimps, Kingsley and his team noticed an interesting change close to the gene encoding a protein called the androgen receptor, which responds to the male sex hormone testosterone and directs the development of all sorts of male characteristics. Looking more closely, they discovered that species that still had the region – chimps and mice in this case – have spiny penises, but humans lack both the DNA sequence and the knob-spikes. Using similar tests to Bob Hill's experiments with the Sonic Hedgehog control switch, Kingsley discovered that it could turn on genes in the part of a developing mouse embryo that will become the genitals, suggesting this truly was the penis-spine control switch. It also worked when they put it into cells taken from human foreskins, suggesting that men still have the biological equipment necessary to make the spines, but merely lack the genetic instructions to do so.

As with the cat thumbs and fish hips, the loss of a small region of our DNA just 5,000 DNA letters long somewhere back in our species' history has made a major difference to men's genitals. To think about it as an evolutionary just-so story, perhaps the lack of uncomfortable penile knobbles made a female more likely to come back for some more loving. This could have encouraged monogamous pair bonding and more baby-making, helping to spread the mutation through the population. There's no hard evidence to prove that this is truly the case, but you may wish to ponder this tale the next time you're in close proximity to a penis. I know I will.

※

These examples highlight the fact that it's risky to mess about with the protein-coding part of a gene, especially if that protein has multiple roles in building a baby. Sonic Hedgehog orchestrates brain cells, fingers and more; Kitlg makes blood, sperm and melanin; Pitx1 does glands and hips; and the androgen receptor is much, much more than a dick-spike generator. Accidental mutations deleting any of these genes at a stroke, or altering the resulting protein in such a way that it can't work properly, cause profound and often lethal effects that are unlikely to be passed on through the generations. But start tinkering with the switches that turn them on or off in different tissues, and a world of new shapes and colours opens up. These elements are evolution's playground, creating big changes with small genetic tweaks.

So that's penises and blondes dealt with. But what about the rest?

CHAPTER EIGHT

Mice and Men and Mole Rats, Oh My!

‘So just how much of a chimp am I?’ It's a question I've asked several politely bemused geneticists over the past couple of years, and their answer usually comes out as somewhere around 98 to 99 per cent. On average about one in a hundred DNA letters is different between us and our hairier primate cousins, adding up to about 30 million across the whole genome. But the rest of it is pretty much the same.

Back in 1975, a smart young researcher called Mary-Claire King[*] published a paper in the journal *Science*. In it, she used a bunch of calculations to show that the differences between the proteins that chimps and humans are made of are too small to account for the physical differences between our species. Her conclusion was that it must be the way genes are used, not the proteins they encode, that makes us so different from our closest relatives. Her idea is now four decades old, but it's only recently that researchers have had the sequencing technology and computing power to start picking it apart on a grand scale. It's not a view shared by everyone – some scientists I've spoken to feel that even small differences in proteins might create important physical differences – but it certainly seems increasingly clear to me and many others in the field that it's the changes in the switches (and also the levels and combinations of the transcription factors that sit on them) that matter.

One big problem with figuring out what's going on here is that it's not clear what these control switches actually look

---

[*] One of the most kick-ass geneticists of her generation, not only did she play a leading role in the discovery of several important genes, but she then used her skills to help identify victims of human rights abuses.

like. Spotting protein-coding genes is relatively easy, as they
have distinctive features and tend to be very similar across
closely related species – a characteristic known as homology,
or sequence conservation. For a long time, people assumed
that important control switches would be the same in different
organisms, too. When I was a graduate student, my colleagues
and I grew slightly obsessed with looking for DNA regions in
between the protein-coding bits that matched up between
humans and mice, with the assumption that they would be the
control switches responsible for turning on and off our
favourite genes. But although some enhancers and other
control elements are extremely similar across large swathes of
the animal kingdom – such as the Sonic Hedgehog switch that
we met in the previous chapter, which is needed to make stuff
at the end of limbs in everything from a fish's fins to our own
fingers – these seem to be the exception rather than the rule.
In fact, the control switches in our genome are remarkably
difficult to distinguish based purely on their DNA sequence.
One way to do it is to look for the characteristic sequences that
transcription factors bind to, although these can be pretty
vague. Another hint is the presence or absence of various
chemical marks associated with gene activity. While none of
these on their own is a definitive answer as to whether a stretch
of DNA can turn a neighbouring gene on, the presence of
several of them together can provide a strong clue.[*]

The 'fallacy of homology' – the assumption that the control
switches for genes will be the same across species – is an easy
trap to fall into, but it doesn't make sense. Surely if all the genes
and all their switches are exactly the same, then different

---

[*]The only way to know for sure is to do experiments: either
removing suspected enhancer regions using genetic engineering
techniques and seeing if the nearby genes still go on at the right time
and in the right place, or – as we saw with David Kingsley's
experiments with his fish-hip switches – by attaching them to a
'reporter' gene and seeing if that gets turned on where and when it's
expected. However, this is a lot of work, so most scientists prefer to
infer the presence of control switches by looking for other markers.

organisms will look the same, rather than different. So why was the idea so pervasive for so long? One person with an answer is Duncan Odom from the Cancer Research UK Cambridge Institute. He comes with a reputation as an opinionated high-flier with a take-no-prisoners attitude and a sideline in fire-juggling. He's also one of the most fashion-conscious of all the researchers I've spoken with (admittedly not a high bar to vault), clad in the kind of stuff sold in punky boutiques in Camden and topped with a back-to-front black cap.

He buys me lunch – a curiously textured and not entirely pleasant vegetable tagine from the institute cafe – and we settle down at a table outside the large glass-fronted building. It's late autumn and there's a chilly breeze. I'm prepared to soldier on but it's Duncan who cracks first. 'Do you want to go inside?' he asks, staring at me intensely. 'Because I'm from Florida originally, so for me it's almost always too cold.' As we head back inside to the noisier but warmer canteen, he tells me about the 'arrogance of the genome', stemming from the glut of information churned out by scientists sequencing anything they could get their hands on. 'The thought was that by comparing DNA from human, mouse, dog, blah blah blah, the regulatory regions would pop out, just like the protein-coding genes.' 'Everyone was assuming that the important bits would be the same?' I ask. 'You're assuming – that's the key phrase – that they're important. But you're only finding what you're looking for. And it turns out that that assumption just falls apart when you actually test it. To this day very big scientists are publishing lots of stories about how important the conserved underlying sequence is, and saying that, because they find the same transcription factor binding sites, then they're functional. But that's bullshit, right? Everything we've done in my lab has butchered that concept.'

According to Mary-Claire King's original idea, now supported by a wealth of DNA data, we're built of pretty much the same stuff as a chimp in terms of the protein-coding genes that we share. So the differences between us and our ape cousins must presumably lie in the switches that turn them on and off. Yet if they're also the same, then you'll get the same thing out. Primate genes plus chimp switches make a chimpanzee, primate genes plus human switches make a human. So why did people assume the DNA sequences of the switches would be the same between different organisms? 'Because they're lazy,' Duncan snaps. 'Sorry, that was harsh. The more forgiving answer is that it was a reasonable guess. I mean, we had such wild success with identifying all the protein-coding regions and explaining the hell out of them that there was a reasonable expectation that the regulatory sequences would at least follow broadly similar rules. Yet they don't.'

To find out what, if any, rules they do follow, he and his team have done an experiment millions of years in the making. They've been hunting for control switches in 20 different species of mammals, from dogs to dolphins, mice to naked mole rats, bats to cats, humans, cows, chimps, Tasmanian devils and more. Sixty-six million years ago, none of these species existed. Dinosaurs ruled the Earth, along with small insect-eating creatures, the ancestors of all mammals, scuttling around trying to avoid the stamp of misplaced dino-feet. Then we got a lucky break. A giant asteroid slammed into the land mass that would later come to be called America, creating the future Gulf of Mexico and wiping out around 80 per cent of all the species on the planet. Among the few survivors were our furry little forebears. Suddenly there was room to grow and diversify, and diversify we did.

Some mammals turned their tastes to flesh, others to plants. Some set their sights on the heavens, weaving webbing between their digits as they evolved into bats. A few tired of life on land and slunk back into the sea to become whales and dolphins. From the same ancestor we get animals as diverse as hefty rhinos, tiny shrews and the cleverest ape of them all, humans. Yet under our varying skins, from leopard-print to whale blubber, all mammals are pretty much the same. We all share a similar body plan, and DNA sequencing has revealed that we have relatively few differences between us in terms of our protein-coding genes. But what about the control switches?

To keep things (relatively) simple, Duncan focused on just one organ – the liver. Unlike something like the brain, which is made up of seemingly endless complicated layers of subtly different nerve cells, liver cells tend to all be boringly similar. This means that there's a good chance of seeing reliable signatures of control switches in samples of liver made up of millions of cells, rather than having to go down to one or a handful of more specialised cells. Our current technology for looking at this kind of thing is good, but it's not *that* good. Also, it's easy to identify a dead mammal's liver, because they all look more or less the same.

First, Duncan and his team had to collect samples of liver from their chosen 20 species. Some were easy. The local hospital, vet school and abattoir came up with the goods for the more common animals. Then he called in some favours from other scientists to get hold of some of the weirder ones. Naked mole rats are fascinating to scientists because they don't seem to get cancer, so there are research colonies of these ugly, hairless rodents (best described as wrinkly sausages with teeth) that can provide tissue samples. Tasmanian devil samples came from Copenhagen Zoo – the only breeding colony in Europe. Getting a bit of whale liver proved tricky until Duncan noticed in the news that a Sei whale had beached itself and died on the Northern Irish coast. A few phone calls later and the UK Cetacean Strandings Investigation Programme – a kind of *CSI Free Willy* – had agreed to give him a cup of liver from the unfortunate animal.

With DNA successfully sequenced from the entire menagerie, Duncan and his team started hunting for the control switches that turn on genes in the liver, known as enhancers. To start with, they scanned through all the DNA looking for characteristic sequences that attract transcription factors known to turn on liver-specific genes. They also searched for regions of DNA wrapped around proteins bearing characteristic chemical 'tags' known as histone modifications (more on this later), which are often associated with active enhancers. It was a huge amount of work, trawling seemingly endless stretches of DNA and comparing them between species, but the results were fascinating. They found that each type of mammal uses many thousands of specific enhancers to turn on genes in its liver. But very few of them shared similar DNA sequences across species: just a handful of the switches turning on liver-specific genes seemed to be strongly similar between animals, even though the nearby genes themselves shared a clear likeness. 'Only about eight per cent of the time do we find a transcription factor sticking in the same place between mouse and human – most of the time it's not in the same location. Now, sometimes it's just moved a couple of thousand bases and it continues to do the same thing, but it's still not the same DNA sequence.'

This would seem to be a fairly solid nail in the idea that we should be looking for similarities between DNA regions in different species to identify control switches. Duncan's work also tells us some interesting things about what's been going on through history. He's found that many control switches, particularly those that seem to turn on genes that are important in specific types of cells (such as the liver), are changing rapidly as mammals continue to evolve. 'Our data tell us that over half of the enhancers in all mammals evolved within the last five or ten million years,' he says, poking at me with a chickpea-studded fork for emphasis. 'That's just staggering – the rate of change is beyond belief.'

In contrast, the genes themselves are resolutely the same. Yet all these diverse mammals have a liver, and a brain, and limbs of some description. How can the genes be the same but the switches become so different? 'From the point of view of the gene, all it cares about is being switched on at a certain level. What it doesn't care about is what's telling it to do that, as long as it gets the correct set of instructions. But there are still a lot of things that puzzle me. In terms of the genome, how do we understand why one transcription factor binding site is occupied, when three doors down there's another site that is a better fit?' 'I guess that makes sense from what we know about evolution,' I suggest, 'Nature doesn't care about how it does it, it just bodges something together.' 'It staggers its way to a solution, more likely,' he scoffs, clattering his cutlery onto the plate. 'The problem is that most people have a crap understanding of the vastness of time.' The problem of conceptualising the passing of time on an evolutionary scale is not unlike an excited pre-schooler waiting for a birthday. 'When's my birthday, Mummy?' 'Not till April, darling.' Wakes up next morning. 'Mummy, is it April yet?' 'Not yet, darling.' Next day. '… How about now?'

Duncan agrees. 'We're human, right? We can't think longer than a couple of centuries. For us the Roman Empire was an eternity ago. But when you look at the genome in people who were alive in the Roman times, which we can do now, it's basically the same gene pool that we have today. That is not enough time for evolution to act. But when you go back to archaic humans, all of a sudden you can look at the genome and see that, oh, we're a lot more divergent now. You can start seeing the accumulation of variation that we have in our current genome.'

Over 60 million years we've gone from being a tiny insect-eater to the dominant mammal on the planet. Yet on a more philosophical level, we can't see the changes that are shaping us right now – how evolution is shuffling our switches to create subtly new forms of humans, or even the inklings of new species that one day may not even be classed as human at all. 'I'm curious about that,' he says, thoughtfully. 'I'm curious about the interaction between technology and selective pressures that shape evolution. Have we changed the balance in a way by using our brain? Has our consciousness changed the rules underlying this game in a way that we don't understand?' 'You mean, have we altered the rules of our own evolution?' I ask. 'We complained about the fact that it was cold outside, so we came into an environment that is 24 degrees and 60 per cent humidity. Only in the last 20,000 years has it been possible to have shelter like that. How does our DNA sort that out? Because all of the signatures that we've found in human DNA so far are ones that were applied by brute-force environmental and competitive pressures.'

He mentions the example of lactose tolerance – the ability to drink milk into adulthood – which has come about because of a small change in the control switch regulating production of a molecule that digests milk sugars. While it may seem to a resolute cheese fan like myself that people who can't tolerate dairy are sadly deficient, in fact it's me that's the mutant. The gene encoding the protein that enables humans to digest milk should get switched off as we move from milk to solid food as infants, as it does in most mammals. But a genetic tweak present in about 80 per cent of people with European ancestry, where dairy farming first started around 10,000 years ago, means that it stays on throughout adult life. In other parts of the world this doesn't happen, and many people from China, Africa and Asia are lactose intolerant. As Duncan sees it, this hand-in-hand evolutionary and cultural change has a big impact on the ability of different populations to thrive. 'If you can drink milk and you can farm cows and be better nourished, then you can out-breed your hunter-gatherer neighbours that can't. Bang! You wipe them out in your area. Things like that, they're easy to grasp and they're perfectly in line with classical evolutionary theory.'

I come back at him with an interesting idea I've heard, suggesting that our ability to manipulate our environment so effectively – along with better healthcare – is actually leading us towards a terrible genetic dystopia. By altering our chances of survival so dramatically, we're keeping variations in the human gene pool that probably aren't all that great. 'Well, I would say there's a flip side to that view. We might be accumulating genetic variation that is slightly harmful, but in a way it's also increasing the genetic diversity of a species. Things can go right as well as go wrong. So for the next global crisis, we could be genetically *better*, ready to deal with it.' 'Readying our genes for when the apocalypse comes?' I joke. 'You laugh,' he says, fixing me with a hard stare, 'but the technical industrial civilisation has an extremely high probability of collapse. If we're thrown back to the Stone Age, what happens? That's not to say we won't have a rebirth of technology – humanity might last a few more million years. But it's a fascinating speculation.' Fascinating or terrifying, we'd better hope we've got something useful in our genomes that is fit for the future.

🐾

Over the past few chapters I've talked a lot about how these switches control when and where genes get turned on – how they come on at the right time and in the right place to build a baby organism as it develops – and how evolution has fiddled with them over time to shape different species. But now I want to delve into a little more detail about what's going on when genes are read.

Open any scientific paper about molecular biology published in the past few decades and you'll probably see a diagram depicting DNA as a long, straight line running across the page. Genes will be shown as thin rectangular boxes sitting along the line, a bit like the carriages of a toy train resting on a track. The start of the gene might be another little box at the front, usually with a big arrow denoting the direction in which it's read. Any control switches are drawn as yet more little boxes, some distance away from the gene itself. Sometimes they're to the left, other times to the right, and sometimes right slap bang in the middle of the gene, depending on the organisation of that particular stretch of DNA. And then there are the proteins. Circles, ovals and other oddly shaped blobs all piling up on top to show where different transcription factors bind to the DNA in order to flip the switches and turn the gene on or off.

This kind of two-dimensional, linear portrayal of genes is ubiquitous in journals, textbooks and the nerdier end of the popular science spectrum. It's also, in my not-so-humble opinion, deeply wrong. Every single cell in your body contains more than 2 metres (6½ feet) of DNA. Yes, you read that right. Two metres. It's absolutely staggering. Probably one of my favourite science facts, right there. Once you realise that these 2 metres have to be packaged up and shoved into the cell's nucleus – a structure half the width of a human hair – it becomes immediately clear that portraying

DNA and genes as purely two-dimensional linear entities is nonsense. Rather than each chromosome being analogous to an inflexible long stick of raw spaghetti, they're coiled up upon themselves like a tightly packed bowl of ramen noodles writhing in a thick soup of proteins and RNA. And somehow within this mess, genes need to be turned on at the right time in the right cells. They need to pick up the right transcription factors to attract RNA polymerase to come and read them, without accidentally turning on any neighbouring genes. If they're not needed at all – for example, muscle genes in liver cells, or brain genes in skin cells – then they need to be kept out of the way. And then there's all the rubbish and garbage DNA, just hanging around and taking up space.

In order to fit 2 metres (*2 metres!*) of DNA inside the nucleus, nature has devised an extremely clever packing method. Although DNA is very long, it's also very thin. And it's also very twisty, what with being a helix and all. Because of the way it's constructed, DNA doesn't just flop about. It likes to twist around on itself, and this means that it also has a tendency to curl around things. Handily, the nucleus is packed full of ball-shaped proteins called histones for the DNA to wrap around. Two turns of DNA – roughly 150 letters in total – are coiled around every histone ball, with a little gap in between. Hundreds of thousands of them are strung along each chromosome like tiny beads on a necklace. This string of beads gets coiled up on itself again and again, the DNA and its histones packing and stacking to make an ever fatter strand. Then there are other kinds of looping and coiling that happen, further organising the DNA according to whether it's useful and needed, or just getting in the way.

It's an incredibly efficient feat of packing, helped out by various proteins that twist up and organise the strands. There's also an array of molecular machines that step in to untwist and detangle the DNA when it gets hopelessly snarled up. We'll come back to this packaging and how it affects gene activity later on. But for now it's time to sit back, put on those weird 3-D glasses, and start thinking about genes in space.

CHAPTER NINE

Party Town

'Sometimes I lie in the bath and imagine what it must all look like at a microscopic level,' Wendy Bickmore tells me, leaning back in her chair as if sinking into a pillow of bubbles. 'This is the forefront of what we don't understand. It has to be three-dimensional, but what dictates it and what role it has is unclear.' It was Wendy's beautiful images of DNA daubed with fluorescent dyes, revealing individual chromosomes and genes inside cells, that first drew me to her work when I was a new and eager grad student. Over the years, she's been painting a picture – literally – of how our DNA is organised inside the nucleus, spying down the microscope on the secretive liaisons of control switches and the genes they regulate. When I visit, she's full of excitement at having just been promoted to director of her research institute, the Medical Research Council's Human Genetics Unit in Edinburgh. The builders haven't yet finished her swanky new office, so we drink tea and catch up in a temporary abode cluttered with piles of paper and dead computers.

Wendy and her team have been using a technique called DNA painting to reveal where each one of our chromosomes likes to hang out in the nucleus. It works pretty much how it sounds, using chemical 'paints' that are specific to each of the 23 chromosome pairs that fluoresce in various exotic colours under ultraviolet light. Each paint highlights where those two chromosomes are lurking and how they organise themselves. Rather than each chromosome snaking around the ball of the nucleus, messily twisted together with all the others, she's found that they tend to keep themselves to themselves, forming a compact, neat territory. It also seems that chromosomes aren't particularly fussed about whether they hang out with their twin or not. For example, both copies of chromosome 18 don't get bundled together. Instead they tend to maintain their independence and do their own thing.

Just as genes aren't evenly distributed across all our chromosomes, the chromosomes themselves aren't evenly arranged in the nucleus. Wendy explains how it works. 'Chromosomes are largely separated from each other in their territories. Obviously on the surface of each territory they kind of intermingle. Then on top of that, the way they're packed in the bag of the nucleus is not random.' She's found that the more active a chromosome is – the more genes within it that are switched on and actively being read – the closer to the centre of the nucleus it sits. And even within each chromosome, regardless of where it's sitting, the parts of it that have active genes tend to be nearer the centre than the parts that are mostly switched off. This arrangement is altered in different cell types depending on which genes are on or off, enabling that cell to do its particular job. There's clearly something happening in the middle of the nucleus that makes it a hot destination for active genes. Conversely, the outer reaches of the nucleus are a bit of a dead zone. Chromosomes banished out here tend to be relatively inactive, and most of their genes are turned off. This arrangement seems to be very common across many organisms, especially mammals such as humans and mice.

So is sending a gene to the outer reaches of the nucleus enough to switch it off if it's not needed? Wendy's not so sure, describing an experiment her lab recently published using a technique that can drag part of a chromosome from the middle to the outside edge of the nucleus. 'One of the nicest pieces of science we've ever done,' she proudly tells me, unearthing a copy of the paper from one of the piles on her desk. Yet although it did make the gene activity levels drop a little, it wasn't that impressive. 'It's pretty wimpy! It turns them down but they don't really get silenced.' Instead, she thinks that if a gene is hanging out at the edge of the nucleus, even though it might be switched on, it can't really get going. But if it's in the middle, where the action is, it can work with high efficiency. It's not so much an on/off switch, more of a nudge one way or the other.

As we chat, we realise there's a nice metaphor we're developing for the nucleus as a city. All the exciting nightlife – the bars and clubs, the fancy restaurants – are in the middle. There's plenty of stuff a gene needs to get its party started (transcription factors and RNA polymerase, rather than tequila). But if you live on the outer edges, the suburbs of the nucleus, everything is more subdued. There just aren't as many good places to go, and everyone just wants to stay at home and watch TV. 'Yes!' she says, laughing. 'If you're in the middle of the nucleus there's lots of goodies for transcription, but if you're at the edge you'll be struggling to have a good time.' 'There are no bars and you don't really want to go out there?' 'Exactly – it's not a good pickup joint!'

Just as people tend to move to different parts of a city over their lifetime, there have been some intriguing studies looking at how the organisation of genes and chromosomes changes over development. As a stem cell in an early embryo changes into a nerve cell, most of its DNA – around 80 per cent – is fairly fixed in terms of whether it's nearer to the centre or further out. But the other 20 per cent changes its relationship to the edge of the nucleus, either coming in closer to get a piece of the transcription action or heading out to the quieter reaches at the edge to settle down.

This kind of experiment shows that bits of chromosomes can get relocated on a reasonably grand scale. But I want to zoom in on the city centre and find out what's happening there. How do individual genes manage to hook up with their correct control switches, and then procure a supply of transcription factors and RNA polymerase to start being read? Many researchers believe that there has to be a direct physical interaction between a control switch and the gene it activates, based on early work with bacteria. Imagine a stretch of DNA as a piece of string, with an enhancer at one end and its gene at the other; making a loop in the middle of the string is the simplest way of bringing them up against each other. This idea is supported by a huge number of research papers that have been published recently using a technique called chromosome conformation capture, or 3-C.

To do it, you take a bunch of cells in a test tube and chuck in a load of formaldehyde. This is pretty nasty stuff, gluing all the proteins in the nucleus to their neighbours and capturing any DNA they're attached to along the way. Then you break up the DNA into short pieces, stick the ends of these fragments together into circles and read the DNA sequence of whatever you've got. The idea is that if two particular pieces of DNA are attached to the same clump of proteins – like a control switch and a gene – then fragments of both the switch and the gene will end up stuck together in the same circle of DNA. There are variations on this theme – 4-C, 5-C and more – but they all work in basically the same way: glue everything in the nucleus together, and see what bits of DNA are in the same clump.

I've sat through a lot of talks recently from scientists sailing on these high-Cs, presenting their findings as definitive proof that certain sequences 'talk' to each other by DNA looping around to bring them into direct contact. The data certainly look impressively cool – usually shown as weird square diagrams with diagonal lines dashing across them – and there's lots of them. But are they showing what their proponents think they're showing? Ever the sceptic, Wendy isn't entirely convinced. 'The problem is you don't know what you're gluing together or how big it is. There's growing evidence – and we've just published a paper on this – showing that it's hugely indirect, and you're just gluing a whole load of things together in some enormous structure. So the fact that you end up with two bits of DNA sequence located together doesn't mean that they were ever necessarily that close to each other, they just got stuck in the same glue.' To go back to our nucleus-as-city analogy, it's like knowing two people were at the same huge pop concert in the town square and assuming they spent the evening standing next to each other. But what if one was in the mosh pit at the front while the other was right at the back of the crowd by the bar, and they never actually met?

One way to find out what really happened would be if each person was wearing a distinctive brightly coloured hat, and you went up in a helicopter to take an aerial photo. If you saw the hats right next to each other, you'd be more certain that both people were interacting. But if they never met, you would expect their hats to stay at a distance. Moving from imaginary gig-goers and their garish hats to real cells, Wendy's used a similar technique to look at different stretches of DNA in the nucleus, labelling control switches and their associated genes with tiny dots of fluorescent paint and using a high-powered microscope to get an overhead view.

She's found that, in some cases, the gene and its switch definitely seem to make physical contact, as their paint spots are right on top of each other. But in other situations it seems that although they're both in a similar region of the nucleus they're not actually touching, even if other methods say they should be. This discrepancy is disconcerting, to say the least, especially as large research groups have spent huge sums over the past few years using expensive glue-and-sequence techniques to build increasingly detailed maps of how different bits of our DNA are supposedly folding up. In 2014, the US National Institutes of Health awarded a hefty hundred-million-dollar grant to build the ultimate picture of our 3-D genome, so it's clearly the in thing.

'It's quite funny,' Wendy says, eyes twinkling mischievously. 'At the same time our paper came out saying that these high-C techniques are open to interpretation, there was a massive paper in a big journal revealing a high-resolution map of the genome made through this technique that cost millions of dollars. We're here with our two-men-and-a-dog-this-cost-50-pence paper suggesting that they've spent all that money just discovering that formaldehyde can make a bit of a mess in the nucleus.'

While the finer technical points of this debate are being thrashed out in the pages of academic journals and at scientific conferences, researchers are busily working on developing more accurate methods for working out what the hell's going on in the nucleus. Yet there's still the open question of exactly how the control switches and their genes are interacting with each other down at the molecular level. 'There's this simple idea of a loop of DNA between an enhancer and its target gene ...' Wendy's pen swoops over a piece of scrap paper by way of demonstration. 'That's fine, and there probably are some genes that are regulated like that in simple organisms. But in complex genomes like ours, it's starting to look much more complicated.'

She picks the example of our old friend Sonic Hedgehog. It's used all over the place in development to build different structures – bits of the brain, the limbs and so on – and each tissue type has its own set of control switches to turn the gene on at the right time and in the right place. But how does a cell manage to make just the right loop to bring the correct switch into place in the appropriate cell type? Nobody knows. And the switch for the limbs is mixed up right in the middle of another housekeeping gene that needs to be turned on all the time in all cells. But how does that work if the switch has to be looped over and connected to Sonic Hedgehog, without interfering with the housekeeper? Again, nobody knows.

Also, DNA is a big, complex molecule wrapped up in lots of proteins. It's an unwieldy thing that takes a lot of energy to bend and flex into specific conformations. If a gene is a long way from its switches, it's hard to see how it manages to find them in amongst everything else that's happening around it. And if the switch is nearby but not right next to it, then it might not have enough flexibility to loop around – imagine trying to bend a thick vacuum cleaner hose so that two points just a few centimetres (a couple of inches) apart are directly touching. It physically can't be done.

There's yet another problem with the looping idea. DNA doesn't just sit there in the nucleus, absolutely motionless. Instead, based on how things appear down the microscope in living cells, it's constantly on the move, writhing and wriggling like a nest of snakes. 'In one cell it might look like this ...' Wendy scrawls a random squiggle on the paper. 'And in another it might look like this ...' Scribble, scribble. 'And it's moving around within that space. This kind of dynamic movement isn't very compatible with the idea of organised loops.'

So what's going on? How are the control switches managing to talk to their genes if they don't physically move over to meet them? Instead of a single elegant loop, Wendy thinks the situation is a lot more disorganised. Rather than being right next to each other and physically communicating, the gene and the control switch just need to be within shouting distance, tangled up together in the same writhing mess of DNA. 'Enhancers are just platforms for transcription-factor proteins,' she explains, dragging her pen over the paper again to draw a squiggly, messy ball – like a single noodle dumped in a heap on a plate – to represent a stretch of DNA. Then she draws a little box on a bit of the line to denote a control switch, and another longer box elsewhere, representing a gene. 'The switches attract high concentrations of transcription factors and other goodies, and that attracts RNA polymerase to the general area,' she adds, putting little blobs and circles into the mess. The idea is that at some point, RNA polymerase will randomly but inevitably stumble on the start of the gene and begin to read it. 'These are very simple principles of physical chemistry,' she says, tapping at the whorl on the paper with the tip of her biro. 'It's just mass action. Concentrate enough stuff in a small space and it will happen through chemistry, so you don't have to invoke fancy structures like loops. And the more stuff you concentrate in there, the more you drive the reaction and the more active the gene is.'

She draws in more and more blobs. It's a bit chaotic by now, but I get the picture. Blobs rather than loops. Random collisions rather than precise interactions. 'I was talking about this idea with a colleague and he started calling them "Bickmore spheres", but I think it's now Bickmore's baubles.' 'How would you like to be immortalised in my book?' I ask. 'Spheres or baubles? Bickmore's blobs? Bickmore's balls?' 'Ummmm …' Baubles it is, then.

This model explains a lot of things about gene regulation that were previously a mystery. For a start, it suggests a reason why genes tend to be more active in the middle of the nucleus: there's more transcriptional stuff available there, and it's easier to get things going. However, transcription factors and RNA polymerase are in shorter supply round the edges. As a result, gene activity levels are correspondingly lower as the chances of these molecules bumping into each other are slimmer. Some people are uncomfortable with the idea that the activity of our genes is determined by such seemingly flaky processes, but the more we know about the fuzziness around the edges of biology, the more an idea like this starts to make sense.

It's also appealing in evolutionary terms. As we saw with the Hemingway cats and the spiny sticklebacks, nature tends to fiddle with the settings in our genetic control switches rather than risk messing about with the protein-coding genes themselves. If DNA had to loop into a particular conformation for the switches to access their genes directly, this would put a number of constraints on the kind of positions enhancers could appear in. But when we look at our genome, and those of other organisms, we don't see evidence of this happening. Instead – as Duncan Odom found with his menagerie in Chapter 8 – new control switches evolve fast, in different positions in different species. Hard to explain with loops, but easy with Bickmore's baubles.

The only way to know for sure would be to zoom right into the nucleus and take an intimate took at the DNA and proteins. The fluorescent paint spots that Wendy and her team use are getting them close, but they need to be perfected in living cells where genes can be tracked as they switch on and off. At the moment they work best in cells that have been chemically 'frozen', giving a single snapshot in time. 'There's a lot of money going into imaging at the moment to try and answer some of these questions – seeing is believing, after all!' she laughs. 'We'd like to look at the DNA in real time, and would also like to be able to look at a specific bit of the genome and find out exactly what stuff has been recruited there. We can't do that at the moment. I think the microscopy is compelling but I've got a completely open mind about how these switches work.'

Whatever they are and however they function, the control switches in our genome are a major part of the story when it comes to understanding how our genes get switched on at the right time and in the right place. Some people, like Mark Ptashne in his art-filled New York apartment, would say that they're pretty much the whole story. But others argue that another element – the switches on top of the switches – also plays an important role. In many ways, this is the most mysterious (and most misrepresented) aspect of how our genes work. Let's delve into the world of epigenetics.

CHAPTER TEN

Pimp My Genome

Epigenetics. In the words of Inigo Montoya in the film *The Princess Bride*, many people keep using that word, but it does not mean what they think it means. Over the past few decades the concept of epigenetics has caught on in the scientific world like a particularly aggressive rash. It's even nudging its way into the public consciousness thanks to breathless articles warning of the epigenetic effects of everything from stress to sunshine, exhorting us to pimp our genomes by drinking green tea or munching broccoli. References to epigenetics have leaked into newspapers, seeped into medicine, and contaminated fields such as psychology and even sociology. More alarmingly, purveyors of pseudoscience are jumping on the epigenetic bandwagon, and the word seems to be used increasingly in the same way that certain people bandy about the term 'quantum' as a hand-waving non-explanation for mysterious things they don't really understand. As a scientist, I worked on epigenetics *before* it was cool, and I find this infuriating. So, what is 'epigenetics' all about?

To put it simply, all your cells have the same set of DNA, but they use their genes in varying ways depending on what they need to do – to be brain, skin, liver, whatever. Because all the instructions are the same, the task of turning the right genes on and off is what scientists refer to as 'epigenetic': something over and above the As, Cs, Ts and Gs of the basic genetic code. Our bodies, and therefore the individual cells that compose them, have to respond to changes in the environment around them. And by environment I don't just mean the trees-and-flowers stuff – I mean everything from the biological soup surrounding a single cell within the body to the big wide world. This flexibility enables us to build bones, brains and everything else in the correct way from the single set of instructions in the genome, and respond to the changing conditions around us as we go through life.

At its heart, epigenetics is all about the interaction of nature and nurture. On one side is nature: the genetic information that is hard-wired in the DNA sequences making up our genome. Then there's nurture: the impact of the environment upon how this genetic information gets used. What follows from this is the assumption that epigenetic information is 'written' into our genes by the environment and influences the way they work. And then there is the tantalising idea that these changes might be passed on from parent to child.

You may have guessed from my somewhat grumpy tone that I've got a few issues with all of this. The first is one of definition. In its purest sense – the way that the coiner of the word, Conrad Waddington, intended it in the 1940s – epigenetics is 'the branch of biology which studies the causal interactions between genes and their products which bring the phenotype into being'. In other words, how your genes and the molecules they encode work together to make you, you.

According to people like Mark Ptashne (whom we met in Chapter 5), this is all due to the power of proteins. In his view, everything boils down to the right combination of transcription factors sitting on the right control switches, turning genes on and off as and when they're required. And this accumulation of transcription factors is the result of changes in the environment of the cell or the organism. For example, molecular signals released when you cut your finger lead to the activation of transcription factors that switch on genes involved in skin cell division to patch up the wound. Or in a developmental context, a particular transcription factor combo could have got there as a result of genes previously being turned on in an embryo, making the right transcription factors for the next step. And back, and back, and back, all the way to the egg. No magic extra stuff required. And yet ...

DNA does not writhe naked and free within our cells, and transcription factors are not the only things stuck to it. It's firmly wrapped around histone proteins, in order to keep all 2 metres of it packaged within the tight confines of the nucleus. DNA plus histones, together with a bunch of associated biochemical paraphernalia, is what scientists refer to as chromatin. Remember that word – I'm going to be using it a lot. Over the past few decades – starting in the 1970s and accelerating dramatically since then – it's become clear that histones can be modified with different chemical tags, each of which conveys different meanings. Some of them declare the underlying DNA open for business, encouraging RNA polymerase to get stuck in and start reading. Others are more like a biological 'no entry' sign, attracting proteins that scrunch down DNA even more tightly than usual so it can't be read at all. While some people maintain that it's all about the DNA and the transcription factors that bind to it – a view that can be roughly summarised with the Bill Clinton-esque 'It's all about the sequence, stupid' – those in the chromatin camp argue that these modifications are just as, if not more, important. After all, how can a transcription factor bind to the control switches that turn on a gene if the DNA is completely inaccessible?

Patterns of histone modifications change as cells respond to messages coming in from the outside world – whether that's their immediate neighbours, cells further afield or the wider environment. Turn a gene on, and you'll see histone modifications turn up that are associated with open, accessible DNA and active transcription. Look at a gene that's switched off, and there will be the 'no entry' modifications. There are a staggering number of scientific papers published about these histone modifications every year as researchers map them in ever greater detail at this gene or that control switch. These chemical tags form an important part of a cell's 'memory', reminding it what it's doing and which genes to use.

Then there's something called DNA methylation – a direct modification of one of the letters in DNA, rather than a change to the histone proteins that package it. It happens when special molecules, called methyltransferases, attach a little cluster of atoms known as a methyl group* to the letter C in DNA at various places throughout the genome. But it doesn't just turn up on any old C. It has to be a C next to a G. And, ideally, there needs to be plenty of these CG pairs hanging about together. Although it might seem insignificant, this tiny methyl group makes a big difference to the shape of the underlying DNA sequence, changing its ability to attract proteins. Imagine sticking little extra lumps of plastic onto the bumps on top of Lego bricks – you'll soon discover that the regular bricks no longer fit. Likewise in our cells, there are different types of proteins that stick to methylated or unmethylated DNA. On top of that, there's an elegant interplay between DNA methylation, the proteins that stick to it, and the molecular machinery that writes 'no entry' marks onto the histones packaging genes when genes get turned off.

The discovery that methylated DNA attracts proteins that are associated with silent, inactive genes led to the idea that it acts as some kind of genetic 'off switch'. It's an attractive idea: want to shut a gene up? Put a bunch of methyl groups on it, and this'll draw in the lock-down crew. But, as happens so often in biology, things aren't quite so simple in real life. For a start, it's not entirely clear whether particular chromatin marks like histone modifications and DNA methylation are responsible for orchestrating changes in gene activity – switching them on or off – or are merely a reflection of other processes at work. To draw an analogy, it's like walking up to a shop door, rattling the handle and discovering you can't open it. Swinging against the glass is a sign reading 'closed', in neatly curling script. Thwarted, you try the shop across the street, where an 'open' sign hangs in the window. Pushing at the door, you step right in. To an untrained eye, it might look like the thing that determines if a shop is open or closed is the sign. But the key reason you can or can't get in – if you'll excuse the pun – is whether or not the shopkeeper has locked the door. The

---

*Three hydrogens and a carbon, chemistry fans!

sign is merely a courtesy to customers, providing information about the state of the entrance.

There's a literal fly in the ointment here too: fruit flies only have a teeny, tiny bit of DNA methylation, nematode worms don't really have any either, and yeast has none, as researchers working on these organisms are keen to smugly point out at every opportunity. So whatever DNA methylation is doing, it can't be absolutely fundamental to life. So is DNA methylation actually doing *anything*? Or, like the sign in the shop door, is it just a marker reflecting other changes? To get a handle on what we do (and don't) know about this enigmatic mark, there's only one person I want to talk to.

Late on a January afternoon out of term-time, Edinburgh University's King's Buildings campus is practically deserted. I can't find anyone on reception at the Wellcome Trust Centre for Cell Biology to let me in to the inner research sanctum where I hope to find Adrian Bird, so I purposefully march through the security door behind a passing scientist, looking like I know where I'm going. Eventually I track him down. It's been more than 10 years since I properly spoke with Adrian – the last occasion being a stressful round of meetings with senior researchers in Edinburgh as part of an ill-fated and ultimately futile plan to further my research career by moving north of the border. There's a pleasing symmetry in the fact that I'm now the one to be giving him a grilling.

Through the thin wall of his office I can hear the clicks and beeps of a Geiger counter as someone the other side tests radioactive DNA samples, giving me a faint whiff of nostalgia for my time in the lab. Back then Adrian was Mr Methylation, putting in near-ubiquitous appearances at epigenetics and chromatin research conferences around the world. His story starts back in the 1970s, when he was a hungry young scientist working in a lab in Zurich. At that time, the only thing known about DNA methylation was that there was an occasional weird fifth letter in DNA, in addition to the

A, C, T and G so beloved of molecular biologists. This was methyl-cytosine, or methyl-C: the letter C with that extra little bunch of chemical stuff stuck to it. Nobody knew where it was in the genome, why it was there or what it did. To find out more, Adrian and his labmates were purifying all sorts of different DNA-cutting molecules from bacteria. Known as restriction enzymes, these form a primitive immune system for bugs by chopping up stray bits of infecting viral DNA. Importantly, they only recognise and cut particular DNA sequences, leaving everything else alone.

Adrian tested one of these enzymes on a sample of DNA that had been synthetically generated in the lab, just to see what happened. As might be expected, it cut it to shreds. Next, he tested it on exactly the same sequences of DNA purified straight out of frog cells. Surprisingly, they remained resolutely unsliced. This was puzzling. Why would the molecular scissors cut lab-made DNA just fine, but struggle with identical DNA that had previously been in a living cell? After all, there was another bacterial restriction enzyme that didn't seem to have a problem cutting the same sequence, regardless of where it came from. Eventually it became clear that the difference lay in the tiny chemical difference between the C and methylated C letters in the DNA sequence, that the enzyme recognised. It could cope with unmethylated DNA, but not the methylated stuff.

These two types of biological scissors – one that could cut DNA regardless of methylation and one that only cut unmethylated DNA – gave Adrian an idea. He takes up the story. 'What we did was find a way of mapping methylation, by comparing the patterns of DNA cutting by each of these enzymes,' he tells me. 'Then the question was, where is it? So the next phase was to look at invertebrates like sea urchins. There you find chunks of methylated and unmethylated DNA, about fifty-fifty in the genome. But if you look at vertebrates, like mice, frogs or humans – which are more interesting – it's mostly all methylated.'

When he looked more closely, he saw that instead of every single C being marked with a methyl tag, there were small chunks of these genomes that were unmethylated. Peering even further, he noticed that they tended to contain long runs of alternating C and G letters – 10 times more than anywhere else in the genome. This is exactly the sequence that is the usual fodder for DNA methylation, yet they were distinctly unsullied by these marks. 'We found these little unmethylated islands amongst the sea of methylation,' he says, rippling his hands in the air like an ocean wave. 'At the same time people were finding C- and G-rich DNA sequences near the starts of genes, so it all kind of coalesced together.'

Years later, the picture of the island paradise in our genomes is clearer. Much of our genome is methylated at CG pairs, for reasons that aren't entirely clear. Because methylation is associated with proteins that lock down DNA in an inactive, inaccessible state, it might be a defence mechanism against viruses trying to get into or out of our DNA, as well as playing a role in controlling rogue jumping genes that litter our genome (see Chapter 16). Or it might just be a generally suppressive thing, quelling errant attempts by RNA polymerase to read random bits of DNA. What we do know is that unmethylated islands are havens for certain proteins that love to flock there. In turn, they attract histone-modifying machines, which add on marks specific for genes that can be switched on. In this way, the unmethylated islands mark out critical regions near the start of genes that are ripe for activity.

The important thing to notice here is that it's the *absence* of DNA methylation that's important, not its presence. This is a slightly hard concept for many people to get their heads around, as we're so used to seeing the *presence* of something as being the meaningful thing. But like the classic optical illusion of a white vase popping into existence between two dark faces, it's the negative space – the absence of the mark – that's key.

So we know that an unmethylated island near a gene is a sign that it's turned on. Then surely it must follow that if it somehow becomes methylated, then the gene will be locked down and switched off, right? Wrong. 'The prevailing view – that there's some kind of switch and when it's unmethylated the gene is on and when it's methylated the gene gets turned off – is not supported by much evidence,' Adrian points out. 'There are a few spectacular exceptions,[*] but actually it's pretty clear that hardly any of these islands get methylated in the cells in the body as we develop and grow in the womb.' One example he gives me is a gene called Alpha Globin, which is only switched on in blood cells. As might be expected, its personal CG island is unmethylated in blood cells. But it's also unmethylated everywhere else in the body, even though it's completely silent. There's evidence that adding DNA methylation might be important for shutting down a handful of genes specifically in developing egg and sperm cells, but that seems to be about it.

Some scientists are now finding evidence that subtle changes in methylation on the 'shores' of these islands, or at genetic control switches further afield, might be more important for controlling gene activity, but it's still quite up in the air at the moment. 'It's a myth that got around,' Adrian sighs, 'that you change the methylation and somehow genes get switched on and off. But it doesn't seem to play a role in normal development.' 'But what about when things go wrong?' I ask, thinking of the pop-science articles that lay the blame for all sorts of ills at the door of dodgy DNA methylation. Practically every week I see new research papers coming out describing aberrant DNA methylation patterns near genes

---

[*]These exceptions include imprinted genes, which we'll meet in Chapter 20. The other prime example is the inactive member of the pair of X chromosomes in female mammalian cells, which gets almost completely shut down in order to compensate for the double X chromosome dosage in females, who have two compared to males who have just one. The whole inactive X chromosome is methylated all over the place, with a load of other chromatin modifications too. And it's consistently inherited from cell to cell as they divide, only being wiped out when a female makes egg cells, providing a clean slate for the next generation. All but the most grumpy of purists would argue that this is definitely an epigenetic phenomenon.

that are allegedly important in all kinds of tumours. What's more, these rogue methylation marks are usually associated with inactivated genes. It's hard to know exactly what role these changes are playing, however. Are they actively involved in shutting a gene down, or are they merely locking a gene into a silent state, after some other event has switched it off? 'Well, you do get odd methylation of these islands in cancer cells,' he admits. 'Is it an off switch? It doesn't really matter, because if you remove DNA methylation using a drug called azacytidine, you turn the genes back on,' he says, slightly dodging the question. 'Regardless of how they got switched off, it might be therapeutically interesting in cancer.'

If you're a cancer patient, you don't care whether changes in methylation were a cause or a consequence of changes in gene activity – all you care about is a cure.[*] Some doctors are now using azacytidine to treat leukaemia and other blood diseases, mostly in an experimental trial context, and there's a huge amount of interest in drugs that can influence histone modifications. But although the results are promising in some cases, it's not a magic bullet by any means. 'It does seem to be clinically beneficial,' Adrian muses, 'so I think the evidence that DNA methylation is perhaps an Achilles' heel of cancer to some extent is sort of supported, without being the whole answer.'

It's a typically equivocal response, and the kind of thing that I've heard from several scientists I've spoken to about the weird world of epigenetics. By now you've probably noticed that I'm drawn to curmudgeons and sceptics rather than florid hype machines, and I have to confess that their cautious thinking has rubbed off on me. Every month there are research papers pouring into the scientific literature looking at DNA methylation in all kinds of contexts, from development to disease. As an example, there was a high-profile paper in early 2015 suggesting that a reduction in DNA methylation in certain parts of a developing male mouse's brain might be responsible for controlling genes

[*]The latest research suggests that demethylating drugs actually work by fooling cancer cells into thinking they've been infected by a virus, rather than mucking about with individual genes.

imparting more masculine gender characteristics. It's the kind of story that hits all the right journalistic buttons: feminine fancies are encoded in a scattered handful of chemical tags affecting certain key genes, while blokeish behaviour is due to missing methylation marks. But there's a long way to go to fully unpick what this means for mice, let alone humans, in the context of normal development. Changing patterns of DNA methylation may yet turn out to have some kind of effect on controlling genes as part of normal life. Yet the data are far from definitive at the moment, and it's certainly not simple.

For all the research papers being churned out, there's still a huge amount we don't know. In some ways there's a lot of stamp-collecting going on – endless mapping of modifications and methylation – without a lot of thought going into whether they're actually *doing* anything. There are a few more interesting twists too. For example, it may be that what look like changes in methylation are actually a reflection of underlying variations and alterations in the DNA sequence itself. Methylation can only be put onto pairs of DNA letters consisting of a C followed by a G – switch either of those for any other letter and the methylating machinery in the cells can't add the tag.

Another problem is resolution. At the moment most techniques to analyse DNA and its methylation still need a reasonably sized lump of tissue, so we only ever see a smoothed-out average over hundreds, thousands or even millions of cells rather than the pinpoint picture within each individual cell. There's also a small but growing list of other types of DNA modification – the extent of which is only starting to be unearthed through in-depth sequencing technology – which may or may not be important for controlling genes. We just don't know yet.

An additional issue is whether (and if so, then how) chromatin changes are inherited as cells divide. Although patterns of DNA methylation definitely seem to get duplicated when DNA is copied, there's precious little evidence that the same thing is true of patterns of histone modifications. Cells do 'remember' what they're doing once they've divided, in terms of which genes are turned on or off, but it's still not clear whether this is because of inherited epigenetic modifications, or just that the right transcription factors are hanging around ready to get everything set up again. A recent paper has shown intriguing hints that certain histone marks can be inherited as yeast cells divide (providing you muck about with their chromatin-modifying machinery enough), but the jury is still out as to whether this is going on in more complex organisms.

Then – and here's where things get controversial – there are all the stories about how epigenetic changes might be passed on to the next generation. There are a few famous examples of epigenetic 'memories' of fear or famine that seem to have been passed down from parent to child. Using the latest techniques, researchers are starting to find traces of methylation that persist in the germ cells that will make egg and sperm, in defiance of genetic dogma that says all such marks should be wiped clean.

I burble all these thoughts at Adrian Bird, in search of some sort of state-of-the-nation view of where we are. 'I believe that it's mostly hype,' he says, shaking his head disapprovingly. 'If it's a question of whether I see DNA methylation as an inheritance system all on its own, I don't think it is one. Methylation tends to get passed on when cells divide, but it's far short of the absolute copying of the underlying DNA sequence, so the capacity to remember anything is limited.' But what about the conversation between nature and nurture? People get very excited about the fact that epigenetic marks seem to change in response to alterations in the environment inside and outside ourselves. Are these changes important here, and can they be passed on to the next generation? He thinks not. 'The evidence that the environment can affect the way in which genes work by rewriting the marks on the genome such as DNA methylation and histone modification is, in my opinion, weak. Of course, there are ways in which the environment affects how genes work. If you get a temperature, it changes the way that certain genes act. There are special genes that are activated when the temperature goes up, but this is done by DNA-binding proteins. The chromatin changes just follow along. You can call it epigenetics if you want, but to me it's just environmental responsiveness.'

As Adrian sees it, the key thing that the evangelical epigeneticists want to believe is that chromatin changes are stable in the long term and can be passed on to the next generation. Again, he's deeply sceptical of these claims, at least for now. 'I'm not saying that it is impossible that there are stable changes in the way genes are used that are influenced by the environment. I'm just saying that the evidence at the moment seems to me to be mostly flaky. But there's a very strong desire to interpret it in a positive way, and the journals and the media love it. So the idea of transgenerational epigenetic effects gets a great press, despite the weakness of the data.'

'And what about DNA methylation itself?' I ask, wondering if he really believes in it at all any more. 'I think people are becoming more sceptical about what it might do. That's because originally we thought it did everything, the way people usually do,' he sighs. 'And we know that it is really important for some things, but not everything.' 'I've heard it described as a "lock" that keeps genes turned off,' I put to him. 'Is that true?' 'Like a lot of things in biology, it helps. It's not the single thing, but it's one of the things.' He adopts a candid air. 'To be perfectly honest, when you get down to a molecular level, despite all the activity and despite all the noise, we actually know rather little about how these things really work.'

As he explains to me, people argue vociferously about all the ins and outs of how these changes might work to affect gene activity, but it's still mostly a mystery. Rather than being a padlock that can be firmly and obviously opened and closed, locking genes on or off, he sees epigenetic chromatin changes in a more flexible way. Once transcription factors have managed to get onto a gene and turn it on, epigenetic modifications make it easy for it to stay that way. And once a gene is off and the DNA is tightly wrapped up, it's trickier for them to get things started again. But it's not impossible.

He rummages in a desk drawer. 'Carrots' he declares. I half expect him to pull one out, but he's holding a lens-cleaning cloth for his glasses. 'You can grow a whole carrot from a single carrot cell without any problem,' he says, removing his specs and rubbing them thoughtfully, 'but why the hell can't you do that in animals?' As keen gardeners will know, there are plenty of plants that will happily grow from stray shoots or leaves. But if you cut off a human finger and put it in a pot, you don't get a new person. For a long time it was believed that animal development was a one-way street, from egg to adult. That all changed in the 1960s, when John Gurdon started fiddling about with frogspawn.

Using a fine glass pipette, John carefully sucked the nuclei out of a huge number of fertilised frog eggs, effectively removing all their genetic material. Then he plopped the DNA of other cells into these emptied biological bags – first trying cells from early frog embryos and then adults – and watched what happened. Incredibly, some of these reconstructed eggs grew into wriggling little tadpoles and transformed into healthy adult frogs, flopping aimlessly in their tank.* This proved without a doubt that the developmental clock can be wound back. Adult cells can be persuaded to forget all the decisions they've ever made and start again. But that's frogs, and they're a bit weird. Surely, many people thought, this can't possibly work in mammals?

Three decades later, in 1996, scientists at the Roslin Institute in Edinburgh led by Ian Wilmut and Keith Campbell proved the cynics wrong when they announced the birth of Dolly the Sheep. She was made by taking an adult sheep's breast cell and putting it into a fertilised egg whose own DNA had been removed. Her birth proved that whatever epigenetic marks are put on the genome in various cell types as they make their journey through life, they aren't irreversible. And the fact that Dolly eventually had happy little lambs of

---

*Rather than the spritely little amphibians you might see around British ponds, these were African clawed frogs, known in Latin as *Xenopus laevis*. Possibly the least exciting lab animals ever, they just hang in the water like bored teenagers at a family party. And they smell just as peculiar too.

her own proved that there weren't any particularly horrendous transgenerational effects from making babies in this way.

Since then, researchers have taken up the cloning baton and run with it, using variations on the nuclear transfer technique first perfected by Gurdon. The cloned menagerie slowly grew: cats, ferrets, dogs, even rare and extinct creatures. But all of them involve putting a donor cell nucleus into an egg, and eggs are pretty special. The assumption was that the egg was full of chromatin remodellers, capable of stripping off the epigenetic baggage the cell had picked up during its lifetime and wiping the slate clean. That all changed in 2006, when Shinya Yamanaka discovered that adding a combination of four molecules could convert adult cells into early embryonic ones. This is effectively cloning in a Petri dish, bypassing the need to go into an egg, and these so-called induced pluripotent stem cells can do everything regular embryo cells can do. But, crucially, the cocktail of molecules needed to turn them back are all transcription factors – nothing more, nothing less. No histone taggers, no DNA methylation removers. Just the same kind of DNA-binding transcription factors that we've already encountered, sitting on control switches and turning genes on. This is fairly hefty evidence that transcription factors are probably at the heart of controlling patterns of gene activity, rather than epigenetic modifications.

Along with a growing number of scientists who've come to realise the hype and bluster around the near-mythical status of epigenetics doesn't quite stack up, Mr Methylation himself is settling in the sceptical camp. 'There is a lot of bullshit about DNA methylation. I think our knowledge of gene activity is actually much less than we would like it to be, but there's no doubt in my mind that the transcription factors are key. Epigenetic systems are absolutely essential and in some cases – such as X chromosome inactivation[*] – we have worked out why.'

[*]See the footnote on p. 108.

What he's less convinced about are some of the data around behaviour. 'There are people who believe that memory is due to DNA methylation. There are people who have published papers that say the way you treat your infants is important for methylation. One of the most cited papers in the whole of epigenetics is about rats that have not been nurtured properly. They do clearly grow up to be more neurotic than rats that have been nurtured properly – this I don't dispute.' What he does have a problem with is the proposed idea for how it's meant to work. According to the researchers who did the study, mother rats somehow create a change in the epigenetic marks at a particular gene in their pups when they lick them, subsequently influencing how they respond to stressful situations. The same team then went on to look at people who'd died by suicide, who had been abused as children. Again, they found changes in epigenetic marks at the same gene, matching the rat story.

Adrian, however, is unconvinced. 'The data are all over the place and I'm deeply sceptical of their conclusion. But this is something that affects humans, and this is the sort of reason why the cult of epigenetics – that this is nurture's way of telling nature what to do – is so insidious.' 'So maybe not an epigenetic revolution quite yet?' 'No. But there has been a *genetic* revolution and we're still reeling from the effects of that. People say, "What did sequencing the human genome actually do for us?" But I think over the next hundred years it's going to do an enormous amount. People expected it to do something over five years, and that's totally unrealistic. We're in a new era in biology, and it's as exciting as it has ever been.' With these sentiments rolling round my head, I head back out into the cold to catch a cab back to Waverley station, tiny flurries of snow drifting down from the darkening sky.

❧

Over the past couple of decades, the word 'epigenetic' has come to mean different things to different people. The purists insist that epigenetic marks or messages affecting gene activity must be passed on from one cell to its daughters as it divides, or (in certain definitions) from parent to child. On the other side, 'epigenetics' now seems to refer to anything that isn't directly encoded within DNA itself – namely chromatin changes such as histone modifications or DNA methylation – regardless of heritability or even whether it actually affects gene activity. The conflation of these two different views is where things get interesting, not to mention controversial.

The eighteenth-century French naturalist Jean-Baptiste Lamarck believed that characteristics acquired in an animal's lifetime could be passed on to their offspring – recall the evolutionary just-so stories of the burly blacksmith hammering away in his forge and passing on his butch muscles to his sons, or elegant giraffes stretching their necks by reaching for the tastiest, loftiest leaves and producing ever spindlier babies. Unfortunately – as proved by the noble sacrifice of countless mice that had their tails chopped off in the quest to prove Lamarck right, only to stubbornly fail to produce shorter-tailed offspring – real life doesn't work like this. However, there are a few intriguing (*i.e.* deeply weird) examples where this phenomenon – known as transgenerational epigenetic inheritance – seems to happen. Perhaps understandably, these tend to generate a huge amount of attention in the media, spawning endless creationist-baiting headlines asking 'Was Darwin wrong?'* and drawing the ire of more sceptical scientists.

One example turned up a couple of years ago when researchers showed that baby mice could inherit memories about specific scents from their fathers, even though they'd never so much as caught a whiff themselves. Another research group claims to have found evidence that the traumatic experience of being caught up in the 9/11 World Trade Center attacks was epigenetically passed on to children by women who were pregnant at the time. And there are stories suggesting that a mother's diet can influence the weight or

---

*The answer to such questions is invariably 'No'.

metabolism of her grandchildren, by causing epigenetic modifications to the precursors of the next generation's eggs or sperm, growing within her developing baby.

Despite the overblown headlines – and the bold claims lying behind them – there isn't a lot of really good evidence that this is the result of changes to histone tags or DNA methylation, or whether it even genuinely happens in humans. Although some examples do seem to hold up in terms of a real effect, such as measurable changes in stress hormone levels in the 9/11 women and their babies, it's very hard to distinguish the influence of chromatin changes on gene activity from all the other physiological and environmental stuff that's going on during the incredibly complex journey from fertilised egg to independent child. This is important because it matters to people. Parents – and particularly pregnant women – face a deluge of conflicting information and judgemental criticism about how to do the best for their baby. Media voices debate the rights and wrongs of criminalising mothers who drink alcohol, while in other parts of the world this talking shop has moved towards actual legislation, holding women legally accountable for perceived harms their behaviour might have inflicted on their offspring.

Of course, there are known harms of drinking excessively or taking drugs while pregnant (although it's a whole other discussion as to whether criminalising mothers for this misbehaviour is a good idea or not[*]), but the evidence of more subtle epigenetic impacts on offspring is much hazier. Not to mention that most of the work has been done in animals rather than people. What if some overzealous law-maker reads the headlines about how a female rat's behaviour or diet can modify the epigenome of her foetus, and starts drafting a bill dictating what pregnant women can and can't do or eat? This isn't the fifteenth century, when expectant mums were advised to avoid eating hares' heads in case their baby was born with a harelip, skip soft cheese lest their sons be born

---

[*]Spoiler: it's not.

with small penises, or shun fish heads for fear of giving their child a trout pout, according to the gloriously wrong-headed folklore of the time. There's more than enough to be worrying about in pregnancy, what with avoiding soft cheese for the small but real risk of *Listeria* infection rather than tiny genitals, not to mention trying to stay off the booze. So let's give mums a break and make sure that health advice is based on real, human science.

CHAPTER ELEVEN

Cut and Paste

Anything found to be true of E. coli *must also be true of elephants.*

Jacques Monod

So far I've only focused on the twisted helix at the heart of all our cells. Yet DNA – and the stuff stuck to it – is only part of the story when it comes to understanding how our genes work. The molecular biology revolution of the 1960s and early '70s was built on the backs of bugs – microscopic bacteria such as *E. coli* and the even smaller viruses that infect them. As it became clear that the same DNA and protein code was at work in everything from bacteria to elephants, as Jacques Monod neatly put it, researchers came to expect that the underlying processes of gene control would be the same too. But it soon became evident that there were a few problems with this view.

Bacterial genomes are models of genetic efficiency: neat and tidy with barely any junk around their genes. Back in the 1970s, scientists knew exactly how big every gene in *E.coli* was, and could match each one up to an RNA message of identical length produced when that particular gene was read. This also corresponded perfectly with the size of the proteins made from each message. But when they started to look closely at human and other mammalian genes, something didn't add up. The RNA messages in human cells were all way too big, each one much longer than it needed to be to encode the recipe to make a particular protein. Weirdly, these long RNAs are only found in the nucleus of the cell. RNA messages in the rest of the cell are a more sensible length that seems to tally with the size of the proteins they encode. One popular explanation at the time was that the ends of the RNA messages were getting chopped off. But the beginnings and ends of both the long and the short messages seemed to be identical. Maybe the top and tail of each RNA got taken off, an unwanted sequence nibbled away, and then the ends stuck back on? Few were convinced.

The recent discovery that humans only have 20,000 or so protein-coding genes poses a further mystery. There are many, many more than 20,000 different RNA messages made from the genome – probably at least 10 times as many. A lot of these come from the so-called non-coding parts of the genome, including accidentally transcribed junk and garbage DNA, as well as important RNAs that aren't translated into proteins yet still help to control gene activity or play other roles in the cell. And even among the genes that do carry instructions to make proteins, it's not as straightforward as 'one gene makes one RNA makes one protein'. There are many more messages than there are actual genes.

These inconvenient truths were ignored or unknown by the early molecular biologists as they prodded and poked at cells to figure out how genes worked. But by the late 1970s a few curious souls were starting to question the idea that genes in elephants – and everything else other than bacteria – were the straightforward recipes that people believed. One of these was Rich Roberts, who won a share of the Nobel Prize in Physiology or Medicine in 1993. To prepare for our Skype chat I did a bit of reading, and before we get to the science there's one question I just have to ask. 'I heard a rumour you bought a croquet lawn with your Nobel Prize money. Is that true?' 'Yep, it's true,' he says, looking more than a little pleased with himself. 'It sits in front of my house. Wait – I'll show you a picture.' He disappears from view and comes back holding a huge framed poster. 'Can you see this? I was Dr December in the Studmuffins of Science calendar in 1997.* Pretty good, huh?' He's pictured reclining on

_____

*A short-lived enterprise put together by US science radio producer Karen Hopkin, apparently as a device to enable her to meet hot, smart guys.

a deckchair on an immaculate green lawn, dressed in equally immaculate sports whites and accompanied by a couple of croquet mallets adorned with a Santa hat. There's a glass of chilled white wine in his hand, thick silvered hair flopping over his forehead, and – I have to admit this – quite a sexy glint in his eye. Along with his cut-glass English accent, tinted with a hint of East Coast colour, he retains a lot of this charm today.

As a boy growing up in 1940s Bath, young Rich always wanted to be a detective. In the mid-1970s, as a newly minted group leader at the Cold Spring Harbor labs, nestled on the thickly wooded northern coast of New York's Long Island, he finally got his teeth into a major mystery. Realising that the idea that genes worked in exactly the same way in all organisms was just an assumption, he set out to discover what was really going on. His weapon of choice was the common cold, or rather the virus that causes it, known as an adeno-virus. As you'll have experienced if you've ever caught a cold, the virus hijacks the machinery of human cells to read its genes and reproduce itself, launching millions of new viruses into the world via your streaming snot. Rich wanted to know if the start sites (promoters) of the virus genes were similar to those of bacteria, which would prove that they worked in the same way. And, by extension, because adenovirus genes are switched on and functional in human cells, using human biological machinery, it would prove that mammalian genes should be the same as bacterial genes, too. True of *E. coli*, true of adenovirus, true of elephants, as Monod might have put it.

An enthusiastic researcher, Richard Gelinas, had just joined the lab in search of a new project, and this was just the job for him. The key to it was coming up with a clever way of getting hold of the RNA messages produced by the virus when it infected human cells. A couple of years earlier, researchers had discovered odd-shaped molecules attached to the start of each string of RNA. A biochemical trick had been developed that could be used to grab hold of them, like a fisherman hooking a wriggling eel by the mouth. By catching all the virus RNAs in this way, Roberts and Gelinas thought they'd be able to match them back up to the DNA sequence of the virus and map where each gene started. But something was wrong.

When they looked at the results of the experiment, using a technique that displays each message as a radioactive spot on an X-ray film, they expected to see 12 spots, corresponding to the 12 adenovirus genes they were looking for. Instead, there was only one. Rich's first response was to suspect that his new recruit had messed things up. 'You know, if you don't do it yourself you always think someone else has done it wrong. Richard had done the work, so I sent him off to do it again and he got the same result. I said, "Look, why don't you let me do it and I'll show you how to do it properly?"' He smiles wryly at the memory. 'So I did it, got the same result and at that point I finally believed it.'

It was clear that the traditional idea of adenoviruses working like bacteria, with different genes each making a separate message, was not the true picture. So what was going on? 'For a long time we were trying to show that in fact we weren't accidentally picking up just one of the messages,' he explains. Perhaps, they thought, there was some technical reason that only one was being trapped and they were losing everything else. A whole series of painstaking biochemistry experiments showed that wasn't the case. There genuinely seemed to be just the one message. The scientific community, however, was unconvinced. 'Awww, they didn't believe it,' he sighs, batting his hands towards the screen. 'No one believed it. You have to realise that scientists – who you would imagine should be really innovative and open to new ideas – are about as conservative as the average religious type. In a way it's a little bit pathetic, because it shouldn't be that way. But although we didn't know what was going on, we knew that it was very different to what was happening in bacteria and it was going to be important.'

To get to the bottom of the mystery, Roberts and Gelinas had to go further than the test tube. They needed to look at DNA itself. It was here that they encountered another problem. Although there may be 2 metres of DNA in every human cell, it's incredibly thin and can't be seen with a regular microscope. So they used a technique called electron microscopy, which uses beams of electrons to spy on tiny biological molecules such as RNA and DNA. 'I came up with this electron microscope experiment that I thought would show what was going on,' he tells me, 'and that's what we did. Neither Richard nor I were electron microscopists, but we had a couple of people just down the hall who were really good. On a Saturday morning we went down to them and said, "We've got this idea for an experiment, could you do it?" And they said, "Well it's never quite been done that way before but we'll give it a go."

Luckily, it worked. Roberts and Gelinas carefully prepared samples of adenovirus RNA message mixed together with the corresponding virus DNA, in the hope of seeing exactly how the message matched up to the genes. The images from the electron microscope were grainy and confusing – twisted loops of RNA and DNA splayed across a background of blobs, like strands of abandoned spaghetti on a dirty plate – but what they revealed was incredible. Each strand of DNA was partnered by a long string of RNA, lining up together in the stretches where both sequences matched. But in certain places the DNA was looping out. It looked as if there were bits of the RNA message missing, so the DNA had nothing to pair with.

To recreate their Nobel-winning experiment for yourself (and impress a nerdy dinner date), all you need is some pasta. Imagine pairing up two lengths of cooked spaghetti on your plate – one short, representing the virus' RNA, and one long strand imitating its DNA. If you match them up at the top and the bottom, then keep pressing them together from the ends towards their middles, you'll end up with a loop of the longer one – the DNA – sticking out in the centre.

Party tricks aside, what did these loopy results mean? It was a tricky puzzle to solve but eventually Rich figured out what was going on. It's just as true in science as it is elsewhere in life that a picture is worth a thousand words. Rather than trying to mentally convert abstract spots on an X-ray film into the looped reality of RNA and DNA molecules, it's much easier to grasp if you can see an image of what it actually looks like. Unlike the situation in bacteria, where each gene makes a single uninterrupted message, the virus was making one long string of RNA covering all its genes, then cutting it up and pasting certain bits together to make the messages encoding proteins.

Instead of the scepticism that had met Rich's early lab results, the pictures were enough to convince people – most importantly, the reviewers and editors of scientific journals – that it was real. 'Normally when you have an electron microscopy result, you're looking at just a few molecules in a very large population, so to try and generalise from that is difficult,' he explains. 'If there really were millions of molecules in your initial sample and you're just looking at two or three, how do you know you're not looking at the weird ones? But we already *had* all the biochemical evidence to back it up, and as soon as people saw the pictures everybody believed it.' Pre-empting BuzzFeed by several decades, in 1977 Rich published his paper describing the phenomenon in the journal *Cell* with the headline 'An amazing sequence arrangement at the 5' ends of adenovirus 2 messenger RNA.'[*] And then, as he puts it, the rest is history.

We now know that this process of RNA cutting and pasting, more formally known as splicing, happens in nearly all of the protein-coding genes in every organism more complex than bacteria. Humans, fruit flies, plants, earthworms and, yes, elephants – you name it, it splices its RNA. We also know that almost every gene is organised into an array of exons (the bits that carry the instructions for making proteins) interspersed with unwanted filler. These are the introns, which get cut out and discarded, while the exons are glued

---

[*]The fourth base pair will blow your mind! There must have been a fashion for superlatives about RNA in the 1970s – the title of one paper published in *Nature* in 1975 notes the 'bizarre' chemical structure stuck onto the end of RNA messages.

together to make the correct RNA message. 'It's as though you're making a movie,' Rich explains. 'You film a lot of different scenes and then the editor comes along and makes the finished version.' His analogy to movie-making is a good one. Once the cameras roll, there's plenty of footage before, during and after takes that isn't actually part of the film. All of that gets trimmed out in the edit. Then scenes get spliced together for the theatrical release, the director's cut, the version for TV, the one for aeroplanes, the alternative ending, plenty of DVD extras and the blooper reel … All of them are based on the same footage, they tell the same story and share key scenes, but each one is unique.

This idea of RNA splicing was an elegant solution to the mystery of the 'too long' messages in mammalian cells, and the data were sound. But even before Rich's paper came out there was trouble brewing. Researchers with exciting new discoveries usually go to scientific meetings and conferences to share their results before their work is published. It's a bit like a band touring festivals over the summer so everyone can hear their new songs, then releasing an album in time for Christmas. On the plus side, it leads to fruitful discussions to help iron out problems with interpreting data, plans for new experiments and promises for samples of much-needed lab ingredients such as DNA or cells. But it also provides an opportunity for unscrupulous people to nick your ideas.

Rich may have been the first person to get definitive proof that RNA got cut and pasted together to make different messages, but plenty of other labs had noticed strange results from their experiments in different systems and were oh-so-close to reaching the same conclusion. 'I went up to Harvard and gave a seminar, and within a few days there were half a dozen other groups who had all found splicing in their systems. I remember I went to one meeting and I talked first because I had a big discovery. Then one of the next guys started explaining all his previously mysterious results in terms of splicing, just on the hoof.' It was then that the trouble really started.

'What happened next is that everyone was trying to show that they had actually discovered it first even though they had heard about it from us, because unfortunately that's how many scientists are. So there was a huge rush to downplay our results and say that their findings were more important.' 'You must have been pretty pissed off, right?' 'A little bit, yes,' he shrugs, diplomatically. Then he becomes more candid. 'It was quite annoying. I didn't get a lot of credit for it at the time.' He starts complaining that Phil Sharp, another researcher who discovered splicing in adenovirus around the same time, never mentioned his and Richard Gelinas's work in subsequent papers – not exactly a crime, but certainly not good etiquette. Disillusioned with the field, and following a failed attempt to recreate the cut-and-paste reactions of the splicing process in a test tube, Rich and his team drifted away to explore other research avenues. Then, one day in December 1993, at around 11 in the morning, the phone rang.

'It was CNN. They asked me if I'd seen the news. It turned out that the Nobel Prize committee had the wrong phone number for me, so I didn't actually get told I'd won until it was all over the media and everybody knew about it.' 'Did winning a Nobel Prize feel like the validation you'd been waiting for?' 'It felt pretty good – I'd recommend it to everybody,' he grins, cheekily. As is often the case with the Nobel Prizes in science subjects, his award was shared with another person: none other than Phil Sharp. 'Were you OK with that?' I ask. 'He had this postdoc in his lab, Sue Berget, who did the work. She had spotted something odd and knew that something funny was going on, which Phil didn't believe. And I think it was only later that he came to realise that what she'd been telling him all along was correct.'

Before we wrap up our chat I ask him if he has any more good gossip about that time. 'You'll have to read my memoir that I'm working on – I don't want to give everything away.' I'm just about to sign off when he leans into the webcam and whispers, 'Here's a story for ya …' 'Go on then …' 'The best thing about being in Stockholm to pick up the Nobel Prizes was that we had to give a lecture about our work. I gave my talk first and then Phil gave his lecture. We'd just made another discovery that I wanted to speak about instead, but Phil talked about splicing. At the end his daughter – I think she must have been about 18, I've known her since she was a little girl – she came up to me afterwards and said that she liked my lecture better than her dad's.' He winks at me conspiratorially. 'That was good.'

❖

Just as I've finished talking to Rich, an email pops into my inbox. Coincidentally, it's his fellow Nobel laureate's PA, who responds in a peculiarly antiquated way to my request for an interview. 'I could propose the following windows for the desired phone conversation. Mayhap one of these might make a match for you?' Mayhap it might indeed.

Phil Sharp is a man who seems to have had his fingers in almost all the pies of molecular biology over the past 40 years, and I'm keen to hear how our understanding of splicing has developed since its discovery. We speak on my birthday – one of my geekiest presents ever. 'Splicing was a totally stunning, unanticipated discovery,' he tells me over a crackly transatlantic phone line from his office at MIT. 'Then as we started looking at splicing in systems that we could study at the time, we recognised that there was enormous variation in the process. One time the gene would be spliced in one pattern, and then another it would be spliced in another pattern, producing two different proteins in some cases.'

These observations solve the problem I outlined at the start of this chapter – namely, how on earth our cells manage to make so many RNA messages from so few genes. To illustrate what's going on, we need to take a brief diversion into the blogosphere. I love cookery blogs. They're like porn for my tummy, with their tempting photos, folksy stories and free recipes. One of my favourites is called Stonesoup, which provides simple instructions for tasty meals made with just five ingredients. The thing I like best about it is the fact the author, Jules Clancy, provides an assortment of options at the end of every recipe. Don't like lamb kebabs? Try beef. Switch grilled chicken for halloumi cheese to make a vegetarian warm salad. Use cashew nuts to turn a prawn stir-fry vegan. Pour on coconut milk for a dairy-free sauce. And so on. One set of instructions gives me an assortment of different meal options to try, depending on what I've got in the fridge and the nutritional quirks of whoever I'm cooking for. And the same process happening in my kitchen when I'm rustling up dinner is going on in my genes as they're doing their own biological cookery.

A gene isn't an uninterrupted list of instructions for making a protein. The recipe is split into chunks of important genetic text (or exons) that are separated by lengthy paragraphs of waffle and nonsense, known as introns. When the gene is read, the transcription machinery copies out the whole thing, introns and all. Then a large conglomerate of proteins collectively known as the spliceosome sets to work chopping it up, ditching the unwanted bits and gluing the exons back together to make a neatly edited recipe. But, just like my favourite food blog, many genes have a whole bunch of different options that the cell can choose from, depending on what it needs to make. Rather than splicing together every single exon, cells skip and switch different parts of the message to create alternative versions. If a gene has five exons, a cell could make an RNA message with all five in a row (1:2:3:4:5), or any other option, as long as it's still in the right order – 1:2:4 or 1:3:4:5, for example. This process, known as alternative splicing, enables cells to make several different proteins from one gene, just by switching up the combination of exons. It's a clever and efficient way of getting more bang for your genetic buck, so to speak.

But in some cases it's much more than getting two or three messages for the price of one gene, as Phil explains to me. 'It's like getting one thousand for the price of one, some genes are so complex in their splicing! We used to think of one gene making one protein with one function, but there are different versions of a protein encoded by the same gene, and they can have different functions too. We don't have a full book-keeping of all that complexity. Maybe we never will, because every single cell could differ slightly from another cell.'

He tells me about a gene called Dscam, which makes a protein that sits on the surface of a fruit fly's brain cells. From their single Dscam gene, fruit flies make not one, not two, but 38,016 different RNA messages. It's a staggering example of alternative splicing at its finest. There are 38 exons in the *Drosophila* Dscam gene, and for four of those there are a number of alternative versions, any one of which gets selected for a place in the finished RNA message. Moreover, each alternative exon is selected using an independent process, so it's completely a matter of chance which ones get picked. Mixing and matching these optional exons creates literally thousands of permutations, and each cell will produce a selection of different versions at random. All of these make a completely unique array of Dscam proteins to decorate each individual brain cell.

There is, of course, a biological point to this complex system. Because cells can distinguish between their own array of Dscam proteins and those of another, it means that nerve cells only connect with their neighbours as the fruit fly's brain gets wired up during development, and don't accidentally create a short circuit with themselves. It's a beautiful and elegant solution to the problem of building a brain: one single gene provides a distinct identity for every single nerve cell, simply through random chance combinations of exons generated by alternative splicing.

Curiously, although flies and crustaceans use this tactic for brain-building, humans do it differently – probably because we have many more brain cells than a fly or a lobster. Instead, we have a couple of hundred genes that make cell-surface proteins similar to Dscam, each of which can make a couple of hundred different versions. Mixing and matching all these variations gives our brain cells their unique identity, enabling them to wire up.

The story of Dscam provides a neat example of how a single gene can provide excellent biological value for money, creating tens of thousands of variations from just one template. But I'm curious as to how this whole business came about, given that bacteria don't seem to do splicing at all. 'Well, no and yes,' Phil says, telling me about peculiar things in bacterial genomes called 'group two self-splicing introns'. These are parasitic genetic elements, like the transposons that litter our own DNA. They randomly hop into a place in the bacterial genome, then cut themselves out again, and the mechanism they use to do this is a very simple version of the complex splicing process in higher organisms. 'So we do know where it evolved from. But the other question is why does it persist? It comes at a genetic cost, meaning that there have to be DNA sequences in genes to specify the splicing.' This is also a risky strategy – if that information gets mutated, the RNA doesn't get spliced properly and might fail to make a functional protein. According to Phil, around 20 per cent of all human mutations affect splicing in some way – either by affecting the molecular machines responsible for the process, or changing the 'cut-and-paste' points in genes themselves. So are there any advantages to it, other than being able to get more messages from each gene? One obvious benefit is that to construct new genes, all you need to do is to shuffle the pieces.

'Nature does that all the time,' he tells me. 'It takes exons from one gene and duplicates them and puts them in the same gene or another gene. There are lots of proteins that show this kind of evolution and exchange of information. This is something that evolution finds useful, and it's true in plants, it's true in humans, it's true in flies, and it's true in worms.' And, presumably – according to Monsieur Monod – elephants. 'Splicing really is a very broad biological phenomenon, and it's all a part of how our genome works. But if you were an engineer designing biological systems, you would not design alternative splicing,' he laughs. 'It is *really* complex and shifts very subtly between different conditions and different signals coming into cells. It is a remarkably weird step in the middle of the process: you've got a gene, you've got proteins, you've got functions for those proteins. Yet here you have this system that mixes all the sequences up. I think it's the weirdest thing ever described in cell biology.'

As Rich Roberts and Phil Sharp discovered, RNA splicing is gene editing on a grand scale. It's the vicious red pen of nature, hacking out large sections of unnecessary waffle and leaving just the protein-making instructions. But there's also another more subtle – and perhaps even weirder – form of editing that takes place as our genes are read, carefully tweaking the message letter by letter.

CHAPTER TWELVE

Nature's Red Pen

Write drunk, edit sober.[*]

It's early September 2014 and I'm sitting in a cafe in the depths of the Stanford University campus in California. I've got a bit of time to kill before my next interview so I grab a latte, hook my laptop up to the wobbly Wi-Fi and open Facebook. Instead of the usual mix of baby photos, drunken pub poses and stupid listicles, my feed looks like a wet T-shirt competition. There's a non-stop stream of men, women, kids and even dogs dousing themselves in iced water. Celebrities are getting in on the act too, with everyone from Justin Bieber to Bill Gates chucking a bucket over their head. This is the Ice Bucket Challenge – a craze that started in the US and became widely adopted in support of the ALS Association, a charity that supports research into the fatal nerve disease amyotrophic lateral sclerosis. The impact so far has been staggering, raising more than $100 million for their work, with many millions also donated to other causes ranging from cancer research to animal sanctuaries in a kind of philanthropic fall-out. But while everyone seems to know about the challenge, I wonder how many people actually know about the illness itself.

Commonly known as Lou Gehrig's disease in the States and motor neurone disease in the UK, ALS is horrendous. There's no other way to put it. For some unknown reason, the nerve cells – known as motor neurones – in a sufferer's body that normally send signals to their muscles telling them to move just stop working, and start to die. Nerve impulses are as vital

---

[*] *This line, often misattributed to Ernest Hemingway, is based on a quote from the novel* Reuben, Reuben *by Peter de Vries.*

as oxygen for muscle cells, and without stimulation they start to waste away. What follows is a progressive and irreversible decline over months, years or even decades towards paralysis and death. People I love have seen its devastating effects first-hand, and we still know of no cause and no cure after decades of woefully underfunded research.

With this in mind, scientists have long been hunting for molecular clues that may explain the disease's cause and point towards cures. It turns out that at least one potential avenue may lie in a genetic phenomenon first noticed more than 25 years ago, which is almost as mysterious and poorly understood as the illness itself.

❧

In 1987, after 12 years in the US, German scientist Peter Seeburg returned to his native country to set up a lab at the Max Planck Institute for Medical Research at Heidelberg University. I suspect he must have come under quite a cloud, as a full-blown scientific and legal storm was about to break. At University of California San Francisco (UCSF) in the mid-1970s, he and his colleagues had been using genetic engineering techniques to put DNA containing the gene encoding human growth hormone into bacteria. The idea was to create bugs that would pump out growth hormone so it could be easily purified. At the time, children with growth hormone deficiency were treated with replacement hormones extracted from the pituitary glands of human corpses. Being able to grow endless amounts of the hormone in vats of bacteria in a lab sidestepped all the problems associated with this unpleasant means of production.

Sensing a profit-making opportunity, the California-based pharma company Genentech wanted a piece of the action. In 1978 they recruited Seeburg to recreate the growth hormone bugs, but what they really wanted was the DNA containing the gene. Unbeknown to UCSF (and completely against the rules), Seeburg slipped back into his old lab in the dead of night and stole a sample of the precious DNA. Fast forward a few years and thousands of children around the world had been treated with Genentech's replacement human growth hormone drug, Protropin, raking in hundreds of millions of dollars for the company. By the late 1980s, UCSF had realised that they might have been robbed, and in 1990 they filed a multi-million dollar lawsuit against Genentech for a slice of the profits. Nine years later the allegations of dodgy dealings flew when the case finally came to court, including Seeburg's alleged theft and his admission of addiction to cocaine and alcohol at the time, culminating in an out-of-court settlement and subsequent $200 million payout to UCSF.

Perhaps aware of the impending lawsuit, it's easy to understand why Seeburg might have chosen the quaint streets of Heidelberg and the cloistered calm of its ancient university rather than stay in California. He also turned his scientific attention to the brain. It was a timely move, as the worlds of neuroscience and molecular biology were colliding. Researchers everywhere were racing to identify the genes responsible for making special proteins that sit on the surface of nerve cells and control their electrical impulses, known as neurotransmitter receptors. They can be activated by molecules we make in our own bodies, depending on the particular type of receptor – such as adrenaline receptors that trigger the famous 'fight or flight' response – but they can also get turned on or off by plenty of different drugs, legal and illegal.

Hunting down genes and reading their DNA sequence was exactly the kind of thing Seeburg was good at. He soon settled down to work and started churning out scientific papers describing some of the genes encoding different receptors in the brain. One project involved looking at receptors that respond to a tiny molecule called glutamate, the main nerve-triggering chemical in the brain. And it was here that he noticed something strange.

As we've seen, when a gene is read in the cell it gets transcribed into an RNA copy of itself. The useless introns get spliced out and a bit of extra biochemical jiggery-pokery goes on to make sure the message stays stable and can be translated properly by the ribosomes to make a protein. Yet despite this cutting and pasting, you would still expect the basic sequence of the RNA message to be the same as the coding bits of the gene it came from. However, when Seeburg and his team compared the RNA and DNA from a few of the glutamate receptors, something didn't match up. In one particular place in the sequence of glutamate receptor 2, the DNA said A, but the RNA read G.* This wasn't just a random one-off error – it was a consistent change, all the time. And it was an important one, altering the chemical makeup of the receptor and drastically affecting its ability to trigger electrical signals. In his paper in the journal *Cell*, published in October 1991, Seeburg described this phenomenon as 'RNA editing'. It's a nice analogy, capturing the precision change of one letter to another just as a human editor might carefully switch an S to a Z when Americanising a text.

Over the years, researchers have sought to understand the hows and whys of this curious correction. We now know that it's not all-or-nothing. The extent of editing differs in various parts of the brain, from 100 per cent of the RNA messages changed in nerve cells in the main bit of the brain (the cortex) to around 50 per cent of them rewritten in the cerebellum (our 'little brain' at the base of the skull). Mice that have been genetically engineered so that their glutamate receptor can't be edited will die of seizures when they're just three weeks old, so it's clearly important to retain some sort of flexibility. Yet it's also incredibly risky, and this is where the connection with ALS, or motor neurone disease, comes in.

In 2004, a team of Japanese scientists led by Shin Kwak did an incredibly fiddly piece of work. Using a pinpoint-precision laser, they carefully cut between 10 and 20 single motor nerve cells from the spinal cords of five ALS patients and five healthy volunteers – around 150 individual cells in all. They then looked at the glutamate receptor RNA message in each of

---

*It's actually a letter I, short for inosine, but it's read as a G by RNA sequencing techniques.

them. As would be expected, every single message in the cells from the healthy people was edited. But in about half the cells from ALS patients, the editing process wasn't working properly. In some cells, the glutamate receptor RNA was still completely edited, but the others had varying levels of correction, all the way down to zero edits. The consequences of this tiny change are profound.

To understand why, we need to look at how the glutamate receptor works. The nerve impulses that fire around our brains and bodies are transmitted by electrically charged metal particles (things like sodium, potassium and calcium) flowing in and out of nerve cells in a controlled way. And the things that control them are the receptors. When a glutamate molecule locks in to the glutamate receptor on the surface of a nerve cell, it's like a key opening a floodgate, allowing calcium to rush in. If the receptor has been edited, switching a single amino acid building block for a different one, the gate only opens a tiny fraction and a controlled trickle of calcium gets through. But the unedited version opens wide, so a huge tide comes crashing in. The nerve cell can't cope with this calcium flood and eventually dies. Repeating this deadly wave thousands upon thousands of times in a person's motor neurones results in the unstoppable symptoms of ALS.

Importantly, this loss of editing isn't seen in nerve cells from the main part of the brain, strongly suggesting that it's something particular to motor neurones and, by implication, to motor neurone disease. Not only does this discovery point towards a potential underlying biological cause of ALS, but it also opens up an intriguing pathway to a possible treatment for ALS by replacing or enhancing editing activity in the motor neurones. Kwak and his team are now trying to use gene therapy approaches to replenish the editing capabilities of nerve cells. Early results in mice are promising, but still a long way off for desperate patients and their families.

Given that editing is so important – and the consequences of its failure so disastrous – why would nature bother to do it at all? Finding someone who can answer this question is one of the reasons I've come to Stanford. Snapping my laptop shut and stashing it in my backpack, I head across campus. I soon start to suspect that finding the right room might actually be some kind of literal entrance exam, designed to screen out unintelligent visitors. Rather than numbering labs and offices in any sensible way, Stanford has adopted a bizarrely arcane system. All the even-numbered rooms are lined up along both sides of one corridor. No sign of the odds, including the one I'm looking for. I stand for a few minutes staring at the blank wall between numbers 340 and 342, wondering if I need to do some Harry Potter-esque move to make the door to 341 magically appear. Eventually a passing student takes pity and sets me on the right path.

Once I've found his office and apologised for being late, Billy Li is energetic and chatty, constantly smiling and drawing pictures in the air with his fingers. He speaks with a fast, clipped Chinese/American hybrid accent, peppering his sentences with the linguistic tics he's picked up after 15 years in the States. 'It is weird, you know,' he says, when I ask him about the glutamate receptor and ALS. 'We have been telling that story for over 20 years and it's a classic, right? But we just don't have many good examples like that.'

Billy is busy searching for new stories. Originally a graduate of Tsinghua University in Beijing, he came to the US to pursue a career in research. Although his initial work as a PhD student led him to discover a new gene involved in a complicated genetic disorder called Bardet-Biedl syndrome, he became intrigued by RNA editing and switched tack. While there were a handful of interesting examples of editing, such as the example of the glutamate receptor, nobody knew how widespread it actually was. So six years ago, working with DNA wizard and part-time mammoth cloner George Church[*] at Harvard Medical School, he decided to do a systematic search to find out.

---

[*]Seriously. The guy is busy pasting together the damaged DNA from long-dead mammoths to try and build enough of a genome to recreate the species (or at least some kind of giant cold-resistant elephant).

It was an epic piece of work. Billy looked at more than 36,000 places in the genome that he suspected might be a good target for editing, comparing DNA and RNA sequences from seven different types of human tissue ranging from brain to guts. To do this, he (or rather, his computer) had to trawl through a staggering 57 million stretches of letters, but it was worth it. The results of his study took the number of known edited genes from just 13 to more than 200. He thinks that around 70 per cent of the edits that he and his team have found make a change to the resulting protein. That's about 150 messages that aren't making the molecule that nature originally intended, based on the underlying DNA sequence. 'Some of it is just accidents, right? We have data to show that,' he says, tapping on a pile of papers on the desk in front of him. 'But in some cases there's a reason for it, like the glutamate receptor. Examples like that are still very rare in the field, you know, but now you can see that people discover a few sites, and gradually they find more and more.'

As DNA and RNA sequencing technology gets faster and cheaper, the data are going to keep coming. They're also going to solve some of the problems that have dogged the RNA editing field: telling the difference between genuine edits in the RNA message and random genetic variations from different people, as well as the challenge of spotting subtle patterns of editing in different parts of the body. 'You know the human genome database we have is not from a single human, right? It's a mixture of many people. And for the RNA libraries we used to work with, they were often from multiple tissues so you cannot tell which individual it comes from, and you don't have the original DNA from that sample. So it was really a mixture of DNA and RNA sequences, back in the old days. But now with next-generation sequencing we can see individual genomes and individual RNA samples, and we can sequence both DNA and RNA from the same person from a small amount of tissue. We can do that fairly easily in the past three or four years.'

Thanks to new technology and faster computers, Billy and his team can now hunt for edits to an extent they'd never previously dreamed of. But it's all very well to find a single letter change in an RNA – it's another thing altogether to show that it actually does something important. 'That's pretty hard to answer,' he confesses. 'It's a major area for the field, and it's kind of pathetic that we don't know anything about their function. We have some ideas to test, but there are easier questions we can address right now.'

In the case of the glutamate receptor, the effect is very strong. Changing a single letter of its RNA has a dramatic impact, but there are very few examples that are as clear cut. The effects of a single edit in a fraction of a particular tissue in the body – say, in just one part of the brain – might be very subtle and difficult to spot in an individual animal or human. There's also a huge amount of editing going on in non-coding RNA, within introns and other bits that don't carry protein-making instructions. Nobody has a clue what this is doing. And most of the editing that's been found happening to RNA seems to be in various bits of the brain, but the reason for this is also a mystery.

I ask Billy another tough question: why would cells go to all the effort of carefully editing RNA messages in this way, especially in the brain? 'The obvious answer is the brain needs a bigger capacity, a bigger repertoire of what it can do. Editing is just a way to diversify what you've got, right? In some brain regions a particular site has to be fully edited, but in other places only 50 per cent of the messages need to be changed. There might be a reason for that. So you can't hard-wire that change into the genome, because you still need the flexibility.' It's costly, biologically speaking, to go to the effort of duplicating, maintaining and controlling two very slightly different genes that basically do the same thing. Editing a single RNA into two (or more) different versions is a quick, flexible way of changing things up without having to muck about with the underlying genome. We're effectively getting two genes for the price of one. 'Yes, two for the price of one!' he laughs. 'This is such a cheap way of making changes. It can happen almost instantly, and then if they're useful they'll be maintained by evolution.'

It sounds to me like there's a whole load of things about RNA that we simply haven't a clue about. He laughs again, deep creases and dimples forming on his face. 'That's why RNA is much more interesting than DNA, right? DNA is just boring!' He gives me a quick history lesson. 'If you look at the history of RNA we have more and more phenomena, starting with splicing in the 1970s.' Since then we've had the discovery of a whole zoo of different types of RNA (which we'll meet in Chapter 14), and a string of scientific revelations showing that this molecule is much, *much* more than a mere messenger. He counts them out on his fingers. 'Now we have non-coding RNA, editing ... Almost every five years you're going to see something brand new.' 'So what's the new thing now?' 'Ah,' he grins, making a ring with his thumb and forefinger and peering at me through it. 'You should go see Julia over in biochemistry.'

CHAPTER THIRTEEN
Ever-increasing Circles

As soon as I walk into Julia Salzman's office in the Stanford University biochemistry department I'm aware that I'm in the presence of maths. An entire wall of her cluttered office has been pasted over with shiny plastic, creating an enormous whiteboard covered in scrawled equations. Julia talks quickly with brief nervous gestures, all wriggling hands and wiggling curls of frizzy brown hair. 'I guess I started on this because of an unusual path that I took. My undergraduate degree is in math,' she says, pointing quickly at the wall by means of explanation (or possibly apology), 'and my doctorate is in statistics.' Thinking that a life in stats was her dream, she took up a postdoctoral job in a statistics department, but soon realised that molecular biology was more interesting. 'I knew nothing about biology at all, except that I was fascinated by it. So I moved to a lab that was doing next-generation sequencing, looking at all the changes that had happened in cancer cell genomes.'

She and her new labmates were looking for fusion genes. Created out of the genetic chaos that reigns within a cancer cell, they're the result of a cut-and-paste job between two unrelated genes. If one is a potent driver of cancer – normally tightly controlled in healthy cells and rarely switched on – and the other is a more workaday sequence that is active all the time, the resulting hybrid is a disaster for the cell: a potent cancer driver that is permanently switched on. The best-known example is Bcr-Abl, more poetically known as the 'Philadelphia chromosome' after the city in which it was discovered.* Accidental joining of two

---

*The story of Bcr-Abl is a triumph of scientific detective work that transformed the way we think about cancer treatment. Read about it in Jessica Wapner's excellent book, *The Philadelphia Chromosome*.

different chromosomes shoves a gene called Abl, which makes a molecule that gives cells a green light to multiply, up against a region of the genome known as Bcr, which is continually active in immune cells. The outcome is the wild proliferation of rogue immune cells in the bone marrow or, as doctors call it, leukaemia. The discovery of Glivec in the 1990s – a drug that specifically switches off Abl – has utterly transformed the chances of survival for people with certain types of leukaemia and other cancers driven by the Philadelphia chromosome. Given the life-saving (and money-making) potential, scientists around the world are eagerly hunting for more of the same.

During their search, Julia and her colleagues spotted something a bit surprising. Trawling through a database of gene sequences, they spotted a relatively tiny piece of DNA that had been pasted into another gene, creating a fusion. It was just 20,000 letters long, which sounds big but is actually smaller than most of the introns in human genes. Her brain started ticking. If such a small section of DNA could end up in another gene, could sections of genes be copied and fused within themselves?

Out of curiosity, she started scanning through a database of sequences from people with leukaemia. Rather than reading all three billion letters of the whole genome, the researchers who created the data had taken a short cut. They had just sequenced the messenger RNA from the cells, hoping it would give them a picture of the underlying DNA it was copied from. As we saw earlier on, these RNA messages always get spliced, chopping out the unwanted introns and sticking together the useful exons to make the finished 'recipe'. These are always in the same order that they come in the gene: exon one joined to exon two, two stuck to three, and so on. Sometimes a few might get skipped along the way but they're always strictly in numerical order. When Julia looked closely at the data, something didn't look right. 'I found that in thousands of RNAs there were exons out of order. It would normally go three then two – something like that.' Again and again she found handfuls of sections glued together in completely the wrong order. Thinking that this might be a peculiar characteristic of cancer cells she had a look at the same kind of data from healthy people. She saw the same thing: exon three stuck to two, eight to seven, two to one. 'At that point we thought, what could be responsible for this? But a lot of people dismissed it. They were only interested in what was happening in cancer cells, not normal cells. But my mentor said, "This is very interesting, you should get to the bottom of it."

Hooked by the mystery, she set to work. The first thing was to figure out if something called trans-splicing was going on. This is where two different RNAs get cut and pasted together, meaning that their exons can get jumbled up in the process. It was the only explanation Julia could think of that fitted with what she was seeing. But all her tests for trans-splicing came out negative, and she wondered if she should just forget all about it. Yet every time she looked at RNA from cells, these strange mixed-up sequences jumped out at her. Then she realised there might be another explanation. Rather than being like pieces of string, with a defined start and end, could these RNAs be circles?

This wasn't a completely bonkers idea. In 1993, Robin Lovell-Badge at the National Institute for Medical Research in London discovered something unusual about mouse testicles. Specifically, he was looking at RNA made from a gene called Sry, short for 'sex determining region on the Y chromosome'. This is the gene that makes men male. It gets switched on as a baby develops in the womb and directs the construction of testes rather than ovaries – the first step on the road to maleness. Even if you have a Y chromosome, if your Sry gene is faulty or missing then you're going to grow up biologically (although not genetically) female. Strangely, although Sry's most important job is during development, it's actually most active in *adult* male testicles, where you might think it's no longer needed. When Lovell-Badge and his team looked more closely at the sequence of Sry RNA from adult mouse testicles, they saw the same thing that would confound Julia two decades later: the messages were scrambled, with the end coming before the start.

After lots of experiments and head-scratching, they realised that they were dealing with circles. During the process of reading the RNA sequence in the lab, the circles were being cut open in random places. If the cut happened near the start of the gene, then it would look like everything was in the right order when it was read: start, then middle, then end. But if the circle was cut in the middle or nearer the end of the sequence, then everything would seem out of order: end, then start, then middle – exactly the same kinds of patterns that Julia had seen.

In the 1990s this was big news. Nobody had ever found circular RNA before, but for some reason it fell out of scientific fashion. Since then there have been sporadic reports of messages that might be circular. Julia mentions one example from fruit flies – a gene called Muscleblind. 'There's a paper from about 10 years ago showing that its dominant RNA form is circular,' she tells me. 'I've mentioned it three times in my work. No one else in the world has cited it, including the authors of the paper!' 'Kind of like having a weird kid you don't want anyone to know about?' 'Exactly! Until we found that this was so pervasive, people who had noticed it thought that if it's a single gene maybe they made a mistake.'

Just thinking of circles isn't enough to convince the scientific world – you need to have evidence. To do this, Julia returned to her mathematical roots. 'I found a statistical way of testing the data to see if it supported circular RNAs and it was overwhelmingly positive. And every other lab test we've done since then has confirmed it. For example, you can treat RNA with an enzyme that eats straight RNA but not circles. And this has no effect on the scrambled exons.' In her mind, the evidence was growing stronger, but she still had doubts. With every experiment that gives you an unexpected result, there's always a moment when you think, 'Did I do that wrong?'

'There were definitely periods of time when I thought I was losing my mind. I thought it was crazy. As a scientist you're trying to find all the possible ways that this could be wrong, so you need to try and disprove it this way and that way.' By this point, the only person keeping her sane was Peter Wang, a researcher who had agreed to help out with the experiments. 'It was really wonderful to work with him because when I thought I was going crazy Peter would do the same experiment in a different room and get the same result. So either we were going mad together or this is what's really happening. It was independent confirmation that either we were both crazy or we really had found something.'

In 2012, she published a paper detailing her results and took it out into the world, speaking about her findings at scientific conferences. 'When I was first giving talks I expected people to say you've made an error, you made some statistical mistakes. But very few people disagreed with the results.' In fact, many of them were probably kicking themselves that they hadn't thought of it first. Researchers would come up to her afterwards and confess to having noticed the same scrambled exons. But unlike Julia's dogged investigation, they had either assumed it was a glitch in the data and ignored it, or put it down to something weird that they were too busy or uninterested to figure out. When I chatted to Rich Roberts, Nobel Prize-winning discoverer of RNA splicing, he told me that he'd noticed RNA circles back in the late 1970s but hadn't known what to make of them. The technician who was doing the work ended up leaving the lab, and it just never got followed up. This kind of thing happens much more often than you might think in the world of research.

Still, it's one thing to find out that our cells are packed full of circles of RNA. The bigger mystery is, what on earth are they doing? At the moment nobody knows. Although most RNAs that are translated into proteins are long strings, it's possible that the circles can be translated too. So maybe they do make proteins, but not very efficiently. Perhaps tying straight RNAs up into circles is a way of controlling how many messages are available for the cell's protein factories to read. Some people think they might be acting as a kind of molecular sponge, mopping up surplus control proteins or small fragments of RNA that act as 'fine-tuners' for gene activity. A 2015 study revealed a large number of circular RNAs in certain cells in the brain, primarily read from genes involved in making connections between neurons, so maybe they're involved in wiring up our brain cells? Meanwhile, other researchers think they're just noise in the system, putting circular RNAs in the category of genetic junk or even garbage – something that accumulates naturally in cells over time, either doing nothing in particular or potentially interfering with normal processes as we get older.

'That's a major unanswered question,' Julia says, shaking her head. 'What I can say is that it's extraordinarily abundant.' She and her team have found huge amounts of this stuff in several different types of cells, and think that around two per cent of all RNA might be circular – a significant chunk when you consider the tens of thousands of RNAs we already know about. To find so much RNA that has gone unnoticed for decades is mind-boggling. It's the biological equivalent of living in Paris for years and never noticing the Eiffel Tower.

Another unanswered question is how they are made. Unlike the loopy lassos that are created when introns are spliced out of messenger RNA, these circles are perfectly round, with no clear start or end. I ask Julia if she has any clues. 'That's a fascinating area that we're studying actively,' she admits, diplomatically. As I see it, there are a lot of blanks here. 'So you're saying you have no idea how they're made, you have no idea what they're doing, you just know there's a lot of them?' I prod. 'We do know *some* things about them,' she reminds me, defensively. 'People in my lab are working on them and we have a lot of data suggesting mechanisms that are responsible for producing circular RNAs. But my sense is that it will take a good chunk of the scientific community to get to the bottom of things – it's unlikely that there will be a simple cut-and-dried answer.'

I ask her how it feels to be on the edge of this strange new RNA world. In response, she bounces slightly in her chair. 'Exciting! It's great, because every problem doesn't have an answer so there's tons of stuff to work on. But it's a little bit challenging because there are still people who will say, "Why are you working on this? It probably doesn't even do anything!"' On the other hand, there are plenty of scientists all too eager to jump on the latest bandwagon. More and more research groups are turning their interests towards circular RNA, hoping to be the first to discover a definitive function or find out how it's made and controlled. Fiddling nervously with her sleeve, Julia admits the field is in for a bumpy time over the coming years. 'I think it's difficult to be patient and diligent, but it's necessary. I think it's the kind of area where there'll be lots of papers that may be interesting but not 100 per cent true.' 'But that's the fun stuff!' I say. 'People presenting controversial ideas at conferences and then lots of arguing about it in the bar afterwards.' 'Yeah,' she agrees. 'It'll be interesting to see where the field is in, say, five years from now – whether there's a consensus, or whether there's still a lot of contention. Eventually I think someone will come up with a consolidating theory that explains all the findings.'

As I gather together my belongings to head back out into the warm California sunshine, our conversation turns philosophical. The history of biology is littered with stories of discoveries that were right under people's noses. Many people had noticed the evidence for circular RNA and some groups had even published papers on it. But the rest of the scientific world was looking the other way, following different obsessions and fads. I'm reminded of those optical illusions that suddenly flip from being a left-facing rabbit to a right-facing duck, or from a beauty to a crone. Once you've seen it, it's obvious. You can't *not* see it. Julia's background in maths gave her a different perspective on the same data that everyone else had access to, and as soon as she pointed out that these weird RNAs were circles rather than straight lines, everyone could see it for themselves. She tips her head on one side with a thoughtful air as she shows me to the door. 'It makes me wonder what else is out there that we're just not seeing.'

❧

So far we've met RNA in a few of its guises: as spliced, edited messenger RNA, carrying the instructions that tell cells to make proteins, and also as circles doing … whatever the hell it is they're doing. But RNA is far from a merely passive messenger, patiently shuttling instructions from DNA to the protein-making machinery. Just as Julia discovered by doggedly digging into her confusing results until they revealed the presence of circular RNA, one of the major roles of RNA was revealed by looking at seemingly baffling findings from a new direction. And it all started with a quest to make a more purple petunia.

CHAPTER FOURTEEN

Silence of the Genes

By the late 1980s, researchers were excitedly exploring the toolbox that had been opened for them by the molecular biology revolution. Like electronic engineers tinkering with circuits and diodes, biologists were starting to hack the genome, using molecular scissors and glue to cut and paste altered genes back into the DNA of various species. One obvious thing to do was to try and encourage cells to make more of a particular protein by shoving in extra copies of the gene encoding it, just as adding extra light bulbs into a circuit board will boost the illumination.

Down in Tucson at the University of Arizona, Carolyn Napoli and Richard Jorgensen were loading petunia plants up with extra copies of a gene responsible for making purple pigment. It was a reasonable assumption that this would make a purpler plant, but unfortunately biology isn't as logically compliant as silicon chips and metal wires. Rather than a rich violet hue, the flowers came out completely white. No colour at all. Or there were strange patterns – stripes and blocks of pure white within the purple petals. Weirdly, their artificial genes were happily pumping out plenty of messenger RNA, which should have been translated into pigment-making proteins, while the plant's own version of the gene didn't seem to be making any at all. Not only were the extra genes working but not making pigment as they should be, but they had somehow managed to turn off the plant's gene too. Realising that nobody had ever seen anything like this before, the scientists cast around in search of a suitable name to describe what they'd found, settling on 'co-suppression'.

Over in Italy, other researchers were fiddling with fungi – specifically a type of red bread mould known as *Neurospora crassa* – and noticed a similar phenomenon. Adding in extra colour-producing genes seemed to have the opposite effect to what was expected, failing to bring extra colour and also turning off the mould's own colour genes. For want of a better word, they called the phenomenon 'quelling'.

Elsewhere, a curious pair in New York – Su Guo and Kenneth Kemphues – were working with little worms known as *C. elegans* of which we'll hear more in Chapter 17. They were investigating a gene involved in how worm embryos figure out how to build a head or a tail at the correct end. As part of their experiments they were testing what happened if they gave the worms RNA messages that were exactly the same letter code as the normal gene but the opposite letter of each base pair.* This is known as 'anti-sense' RNA. The idea was that a reversed, anti-sense version of an RNA message given to the worms would pair up with the regular message RNA and interfere with it, stopping it being read by the cell's protein-producing factories. And this is indeed what they saw: the anti-sense RNA appeared to knock out the RNA from the gene they were interested in. But then things got strange.

Like all good researchers, they did a control to make sure nothing was awry in their experiments. As well as testing their anti-sense RNA, Guo and Kemphues also tested a normal, right-way-round RNA as a control. This 'sense' RNA should have had no effect as it can't match up with itself. Instead, they saw that it also seemed to switch off the worm gene. This was baffling, to say the least. 'It is not clear what accounts for this effect,' they dryly note in their paper, which is the scientific equivalent of saying, "We haven't got a clue".

❧

One of the people who finally figured out what was going on was Andy Fire. Today he's settled in Stanford University in California, but back in the mid-1990s he was at the University of Washington in Baltimore, working in collaboration with Craig Mello up the coast in Massachusetts.

---

*A pairs with T and C always goes with G, whether it's DNA pairing with DNA, RNA with RNA or DNA with RNA.

Andy's office is starkly bare, devoid of the usual personal and scientific bric-a-brac that tends to litter most researchers' lairs. The exception is a coconut adorned with googly eyes and a top hat, perched inexplicably in the corner. When I ask him about it, he tells me it was a gift from a colleague who insisted he needed *something*. He's quietly geeky, muttering at me through his neat beard rather than expounding grandiosely about his ideas and achievements like so many of the other scientists I've met on my travels over the years.

'It was all pretty frustrating,' Andy tells me. 'The plant and fungus people were doing these experiments adding in genes and getting unusual results. We started looking in worms and found we were getting the same kind of thing, but it would be hit-and-miss. Sometimes it would work, and sometimes it wouldn't.' Co-suppression. Quelling. Sense *and* anti-sense RNAs switching off genes. Whatever did it all mean?

Then Andy and Craig had a bright idea. When Guo and Kemphues made their sense and anti-sense RNAs to test by injecting them into worms, they did it using an RNA polymerase purified from a virus. Although it's pretty reliable at making the right thing from a DNA template, the viral polymerase can sometimes mess up and accidentally read the DNA in the other direction, creating an anti-sense RNA. So maybe this silencing wasn't to do with a single strand of RNA, whether sense or anti-sense, but double-stranded RNA – RNA paired with RNA, in the same way that single strands of DNA pair up to make the double helix, created by accident in a test tube and then unwittingly given to the worms.

With painstaking precision, Andy and Craig set about testing different combinations of single- and double-stranded RNA, sense and anti-sense, and were surprised to find that their hunch was correct. Neither sense nor anti-sense single-stranded RNA on its own was enough to interfere with gene activity. It had to be a matched double-stranded pair, corresponding to the DNA sequence of the gene they were trying to shut down. Furthermore, it didn't matter where they injected the stuff – nose or tail – it shut down gene activity across the whole animal. Amazingly, they only needed to add a teeny tiny bit of it to see a potent effect – just a couple of molecules per cell.

Given how many RNA messages are usually made from an active gene (clue: lots), this was an incredible finding. Either there had to be some kind of colossal amplification, turning a handful of RNA pairs into many thousands so they could personally interfere with all the RNA messages for that gene in the cell, or one double-stranded RNA could interfere with lots of RNA messages, one after the other. Alternatively (or even additionally), Andy and Craig speculated, maybe all it took was a few double-stranded RNA molecules, somehow acting directly on the gene itself to shut down transcription. In fact, we now know that all three of these things can happen.

Their paper, published in *Nature* in 1998, is a great piece of research. It's also a masterful work of scientific understatement, given how revolutionary the phenomenon – then termed RNA interference, and now known as RNAi – turned out to be. In it, they speculate that RNAi 'probably exists for a biological purpose', and might enable scientists to study the function of 'many interesting coding regions'. Their discovery made the news just before I made the transition from undergraduate to grad student, and. I remember everyone talking about it with huge excitement, before scurrying off to the lab to make double-stranded RNAs against their favourite genes to see what they did. All the previous experiments started to make sense, if you'll excuse the pun. The random arrangement of multiple copies of the extra pigment gene stuffed into petunias was accidentally leading to RNAs being made in both directions, creating double-stranded RNA pairs that shut off the plant's own pigment genes. Same thing with the mysterious quelling in the bread mould. And Guo and Kemphues's strange results were explained by the contaminating presence of sense (or anti-sense) RNA in their anti-sense (or sense) tests.

Over the intervening years, more and more parts of the biological machinery that turn a handful of paired RNA molecules into powerful gene silencers have been uncovered. Some of them have wonderful names such as Dicer, Drosha, Pasha and – my personal favourite – Argonaute. At the heart of it all seem to be short little fragments of paired RNA, just over 20 letters long, made by hacking up a longer double-stranded RNA. These fragments then seek out any regular RNA messages that they happen to match. Once in place, they attract the attention of snippy biological scissors that chop the messenger RNA into pieces so it can't be translated to make proteins.

Getting rid of the messages made from a specific gene is a pretty powerful way of damping down its effectiveness. But to make a real impact, it's vital to cut the RNA messages off at the source. This is where the magic happens. As well as getting rid of particular unwanted RNA from cells, the small interfering RNAs somehow manage to 'talk' to the gene these messages came from. Exactly how this happens isn't entirely clear,* but the end result is that transcription gets switched off and the gene is locked down into a silent state, with epigenetic marks like DNA methylation and its associated repressive chromatin baggage turning up (see p. 103).

Short interfering RNAs are so potent that just adding a few of these little fragments to cells or even entire animals is enough to turn off the gene they match. This discovery sparked a booming industry of biotech companies churning out manufactured pairs of tiny RNAs to order, ready to 'knock down' any gene of your choosing. Incredibly, it works in pretty much everything from yeast all the way up to humans. RNAi-based drugs made it into clinical trials with dizzying speed, just six years after the phenomenon was first

---

*One of the best guesses at the moment is that the short RNAs hook up with the regular RNA message as it is being read from the target gene, somehow messing things up and shutting down transcription.

discovered. It's a truly transformative technology that deservedly won Andy and Craig a Nobel Prize in 2006.

While scientists have indeed got busy studying 'many interesting coding regions' with these short interfering RNAs, there's still the other question posed in Andy and Craig's paper. The RNA interference system wasn't designed so that scientists could do fun experiments tinkering with their favourite genes. It's there because cells need it. But for what? 'At first it appeared to be an anti-viral system, from the work that was done in plants,' Andy tells me quietly. Many viruses create double-stranded RNA from their DNA as part of their life cycle, so it's not a giant leap to assume that cells might find a way to attack these pairs as a way to shut down virus genes. We know that this happens in plants, worms and fruit flies. But when it comes to animal cells, including our own, things are less clear.

'There was a debate for a long time about whether this was true in higher organisms, because we have other ways of dealing with long double-stranded RNAs that are much less specific. If your cells see double-stranded RNA, they panic and stop making proteins. They shut everything down, not just one gene. Then after a while they turn everything on again and hope the virus has gone away.' 'But small interfering RNAs still work in human cells,' I say, thinking of the multitude of scientific papers showing the exquisitely sensitive effects of these tiny fragments on individual genes, 'so surely we must have that machinery in us?' 'Ah yes – that was first discovered in fruit flies,' he drawls softly, nodding with recognition. 'Scientists found that if you just took short segments, they don't induce this dramatic response but you still get the specific gene interference. So that suggested that our cells have the capability for doing that.'

The fact that RNAi works on specific genes in all kinds of cells, including our own, tells us that higher organisms must have the molecular bits and bobs they need to do RNA interference. But it still doesn't answer the question of what its 'proper' function might be. 'I think that to some of us working on simpler organisms like worms it was obvious it had to be an anti-viral mechanism, because it made so much sense. But the data weren't there to say that this happened in mammals, and there was no really good example of a virus that it might work on in human cells.'

That may have changed in the past few years, with the publication of a handful of controversial research papers claiming to have found evidence of mouse cells fighting off virus infections with specifically targeted RNA interference. Even so, many researchers are convinced that the RNA interference machinery is hanging around in our cells for a different reason. And as we've seen, evolution tends to take things that are hanging around and run with them. 'If you have a mechanism that was once anti-viral, evolution will co-opt it to regulate genes,' Andy explains. 'And if you have a system that's used to regulate genes, it might be co-opted to give you protection against viruses.' 'So is RNA interference involved in regular gene control in our cells, as well as combating viruses? What's your hunch on this?' 'I think it probably is, and that many of the same mechanisms are involved in both. You know that some of the cells in our immune system are actually used in normal development? I think this RNA interference system is the same thing.' Andy's feeling is that RNA interference acts as the 'immune system' for our genome, helping to fight off invading viruses from attacking our DNA in the same way our white blood cells protect our whole body. But because it's a handy mechanism for turning specific genes off, evolution has probably put it to good use as part of the complex system that turns at least some of our genes on and off at the right time and in the right places.

So far all the RNAs we've been talking about have been perfect matches, with the short fragments exactly pairing up with a corresponding sequence in their target RNA (and, by definition, in the DNA of their target gene). But there are also short RNAs that aren't quite identical, with a letter or two different. 'Ah,' says Andy, grinning, when I ask him about them. 'That gets us into the world of micro-RNAs.

Those are clearly involved in normal cell processes, but we don't really know how.'

❧

Someone who's searching for solutions to this mystery is my friend Eric Miska – the deadpan German we first encountered way back at the beginning of this book. Sitting in his darkened office at the Gurdon Institute in Cambridge, half a world away from sunny Stanford, and bathing in the eerie glow from his bubbling fish tank, he tells me about the micro-RNA world. 'I started getting interested when micro-RNAs had just been discovered,' he says, leaning back in his chair with his feet up on the desk. 'The first one was discovered in worms in 1993, but there wasn't much else back then. Then more were found, and things just exploded.' Boom! He pushes an imaginary cloud of molecules into the air.

Micro-RNAs are so hot right now. They're thought to be ubiquitous in plants and animals, and have been the subject of thousands of scientific papers. Created from short RNA messages encoded in peculiar little genes, they seem to be involved in controlling the levels of gene activity across the whole spectrum of life from cradle to grave, influencing the earliest moments of embryonic development through to cancer and other diseases. And there are thousands of the things. Until recently it was thought that there were nearly 2,000 distinct micro-RNAs made from the human genome. At the beginning of 2015 that number jumped to more than 5,000, after researchers in the US took a closer look. And there may be even more that are yet to be discovered. In fact, estimates suggest that up to two per cent of all genes in worms, flies or mammals are used to make micro-RNAs – a truly staggering proportion.

To all intents and purposes, the genes that encode micro-RNAs look more or less like normal genes, just smaller. They have a proper start and end, and they're read by the same RNA polymerase that transcribes all the rest of our regular, protein-coding genes. What's different is the way the micro-RNA messages from these micro-genes get processed. For a start, there can be instructions for several different micro-RNAs all strung together in a row in a single gene, which are read into one message. This then gets folded and chopped up into the appropriate number of specific 20-or-so letter long double-stranded RNA fragments, similar to the short interfering RNAs that we met earlier. Like those other molecular trouble-makers, micro-RNAs seek out matching RNA messages, targeting them for destruction or preventing the RNA from being translated to make a protein.

Because they're much less picky about which RNA message they pair with, each micro-RNA can have many possible targets. In keeping with this, tests on cells grown in the lab show that mucking about with specific single micro-RNAs alters the activity levels of hundreds of genes. This promiscuity means they could have a hand in virtually every biological process going on in cells. Big stuff for such small molecules. Unfortunately, here's where things get *really* complicated.

With a few exceptions, experiments designed to discover exactly what individual micro-RNAs actually do in cells have been disappointing and confusing, to say the least. Although knocking out the entire machinery required for micro-RNA generation and function is disastrous – 'incompatible with life', as the jargon goes – altering or getting rid of any one particular micro-RNA gene seems to have no discernible effect on development or disease. Eric himself set about systematically removing micro-RNA genes one at a time from worms. He found that only about 10 per cent of them are actually important for keeping the things alive and wriggling, and it's probably a similar proportion in mice and humans. Yet it seems odd that biology would go to the trouble of making so many micro-RNAs if they didn't really do anything. For now, the best idea seems to be that rather than acting as master off switches for specific genes, most micro-RNAs are more like sensitive dials, subtly fine-tuning gene activity levels on a much broader scale. They're just one of many cogs in our molecular machinery, all acting together to make sure that cells do the right thing at the right time.

Eric recently switched his attention from micro-RNAs to another class of non-coding RNA controllers, known as pi-RNAs. Mostly found in the germ cells that become eggs and sperm, they're just a little bit bigger than micro-RNAs. The 'pi' bit stands for 'PIWI-interacting'. In turn, PIWI is the name of a gene more formally known as 'P-element induced wimpy testis'. This rather strange name comes from scientist Haifan Lin, who came across some male fruit flies with unusually tiny testicles while searching for genes involved in making stem cells. Realising that 'wimpy testis' – his first response on seeing the small-balled flies – probably wouldn't make a great name for a gene, he had to come up with a more elegant acronym.

Like micro-RNAs, pi-RNAs have been found all over the place in a multitude of species, including our own. In organisms that don't have DNA methylation, such as worms and fruit flies, they're thought to be responsible for keeping a check on virus-like transposons – 'jumping genes' that can wreak havoc in the genome – stopping them from hopping about in eggs and sperm. But there are intriguing hints that they might also be able to switch off regular genes, too. 'Part of the genome is set aside in many animals to make these pi-RNAs,' Eric explains, 'and they match up with transposons, the jumping genes, to silence them. But there's a neat evolutionary trick how this can work: if one of your proper genes happens to be partly similar to one of these pi-RNAs, for example by a transposon jumping into a normal gene, then this gene can be silenced too.'

Even more intriguingly, Eric tells me, it seems that the repressive effects of pi-RNAs might be passed down several generations – at least in worms, and maybe in fruit flies too. Is this evidence for the strange and elusive transgenerational epigenetic inheritance that some people are so excited about? Or is it just a weird worm thing? 'This is not really that weird,' he says, dismissively. 'Similar things have been well known in plants for many decades, and worms are really half-plants!' I look at him quizzically. Surely worms are animals, and plants are ... well, plants? But down at the molecular level, there are actually lots of similarities – remember that the RNA interference field started with a purple petunia. Of course, worms and plants are not enough for our mammal-centric mindset, and the search is now on to find something similar in human cells. Given the history of discoveries in genetics, it'll probably turn up. As one scientist told me, 'Plants show us what is possible in biology' and it's just a question of finding out how these mechanisms are used by other organisms to their own particular ends.

As if that wasn't all complicated enough, micro-RNAs and pi-RNAs aren't the only players in town. The list of different types of non-coding RNA that have been implicated in controlling gene activity – whether by interfering with messenger RNA or shutting down genes directly – is growing inexorably. Naturally occurring short interfering RNAs have been discovered. RNAs can be read from the control switches in DNA. Long stretches of RNA get churned out from the gaps in between genes, and much more. There are lots of ways to be an RNA.

'To me, it seemed really cute that in addition to the 20,000 to 30,000 protein-coding genes in the genome there might be another thousand non-coding micro-RNA genes that are interesting and have regulatory roles,' Eric tells me. 'It was great to discover this whole gene-control mechanism that had been missed, but really not a big shock to the system, I would say.' He sighs and throws his hands up again, this time in despair rather than exaggeration. 'Fast forward to now, and we are in the situation where it is really unclear how much non-coding RNA is out there that is functional. It is a fantastic RNA zoo!' He laughs heartily, finally cracking more than a wry smile. 'It reminds me of when I started out in life, studying physics. At the time there was this particle zoo, when suddenly people got new machines to play with – these accelerators where they could detect more and more different fundamental particles. Suddenly there seemed to be hundreds of fundamental particles, and it was not beautiful.' Scientists like things to be simple and elegant. A wild proliferation of particles is neither, and eventually the number came down as physicists sorted the subatomic wheat from the chaff. As Eric sees it, the problem is that everywhere researchers look they find new RNA transcripts, each very slightly different from the next. But are many of these exotic RNAs actually doing more or less the same thing?

'People find an RNA and say "Hey, I'm going to give it a cool new name and publish a great paper!"' he complains. 'But we really need to understand how many different classes of non-coding RNAs actually exist. We need to categorise them. In particle physics it happened when researchers realised that these hundreds of elementary particles aren't really elementary, they're all made up of a small number of quarks. And from these essential quarks you could build all these other particles.' What he wants to see is a bit of order – to come to a state where a student could open a textbook (or, more likely, a web page) and find a definitive list of different types of RNA and what they do. He might be a long time waiting, because, as we've seen, the world of RNA is hugely complex. Strings of RNA can be any length, from a few letters to thousands. The molecule can pair with itself, with another RNA, or stick to proteins. It can form beautifully ornate three-dimensional structures, which we currently can't easily predict or interpret from the order of letters alone. All of these properties pose huge challenges to RNA biologists as they try to figure out all the things that this amazing molecule does inside cells. Furthermore – and this is important – it can also recognise specific sequences of DNA.

This raises the fascinating idea that non-coding RNAs might almost be the opposite of the transcription factors that recognise the control switches near genes and activate them – but instead of being the 'on' switch, RNA turns genes off. Like a transcription factor protein homing in on the particular DNA letter phrase of its binding site, RNA has the ability to target any specific DNA sequences it matches up with. It's a really neat idea, but for the moment needs a lot more hard data to prove that it happens on a widespread basis. And there isn't a lot of strong evidence to prove that small RNAs can actually home in on their matching sequences in the genome, suggesting that they might be recruited in a different way. So there's plenty to keep RNA researchers busy for a good while yet.

Finally, Eric gives me another possible explanation for what at least some of this non-coding RNA might be up to: it might actually be directing the manufacture of proteins after all. 'We started off with this idea that genes are protein coding and they're easy to find in the genome, right? Now it turns out that quite a few long non-coding RNAs, maybe between 10 to 40 per cent of them, could potentially be used as messages to make short little proteins.' These are what's known as small open reading frames, or Smorfs. Previously they were ignored because they fell below the arbitrary cut-off point that researchers use to define a 'proper' protein – around a hundred molecular building blocks (animo acids). But biology doesn't care for such artificial definitions.

'So we're going all in a circle now!' he sighs. 'Maybe some non-coding RNAs are proper protein-coding genes that have been missed. It seems to me that there is absolute proof of some, and pretty good computational data for quite a lot more of these Smorfs.' Maybe he's right. Perhaps the function of some of this so-called non-coding RNA is actually to encode proteins after all – they're just dinky ones. And it highlights the problem that scientists face when trying to mine the genome to figure out what's in there and how it works: you only find what you're looking for.

We've talked so long the last of the daylight has faded and we're lit solely by the glow from the fish tank. Eric points at the nifty creatures darting through the bubbles. 'If you've got a fishing net, you're not going to find the fish that are smaller than the holes in it. That seems to definitely be the case for these little proteins, so who knows what else is out there? I think these are revolutionary times. There are a lot of wonderful ideas, a lot of possibilities, so let's see how we're going to sort this out.'

CHAPTER FIFTEEN

Night of the Living Dead

When there's no more room in hell, the dead will walk the earth.
George A. Romero, *Dawn of the Dead*

Zombies are everywhere, invading modern culture with their insatiable lust for human flesh. We have TV shows and movies about zombies. Classic works of fiction have been reworked to include their mournful call of 'Braaaaaiiiinnnnns'. Video games allow players to blast the undead into gory fragments. People gather in major cities for 'zombie walks', splashing themselves with fake blood and imitating the familiar creepy shuffle. There's clearly something inherently fascinating about the idea that the dead can escape the grave and come back to dwell among the living. And without wishing to freak you out too much, you have thousands of dead genes scattered throughout your genome, and some of them are just itching to spring back to life.

More scientifically known as pseudogenes, these are the supposedly long-dead relatives of normal, functional genes. There are more than 10,000 of them in your genome, and most biologists would put them in the category of junk or even garbage DNA. Some of them are just regular genes that have had a terrible accident – some kind of disabling DNA change leaving them unable to work properly, or just killing them dead. In the worst case scenario, the damaged gene can actually cause problems. For example, it might make a short or faulty protein that interferes with the normal workings of the cell. In other cases, pseudogenes are created by mistakes in the DNA copying process. Like accidentally reading the same passage twice in a book because you've lost your place, the molecular machines that copy our DNA can slip backwards and repeat the sequences they've just done.

One protein-coding gene becomes two identical genes, or even more. But because the cell usually only needs one normal, functioning copy of that gene, it doesn't matter what happens to any duplicates. Over time, they pick up mistakes that affect their function, eventually losing their ability to code for proteins and heading towards the genetic graveyard.

There's another way that pseudogenes can be made. When RNA is read from regular genes, the messages get chopped up and processed to put them in the right format to be used by the cell. Unwanted introns are spliced out, the protein-coding exons are glued together, and the RNA is given a special tail-sequence that protects the message and directs it to the right place in the cell. Researchers poring over the early drafts of the human genome were surprised to see a whole bunch of genes that looked like they had already been processed in this way – all exons, no introns and bearing that characteristic tail. Normal genes just don't look like that. What's happened is something called retrotransposition, when the processed RNA message from a gene accidentally gets pasted *back* into the genome in a random location. This is a heinous violation of the official central dogma of genetics, which says that 'DNA makes RNA' (not the other way round). However, real biology doesn't care about such things. It happens – deal with it.

Most of the time these cut-and-pasted genes don't work. Because they popped back into the genome at random, they usually lack the necessary control switches to turn them on – they're effectively dead on arrival at their new location. As a result, pseudogenes have been viewed for a long time as belonging in the graveyard of DNA, of interest only to evolutionary archaeologists who dig about in genomes for clues about the origins of different genes and species. But as techniques for detecting gene activity have become more sensitive, scientists have discovered that hundreds of pseudogenes seem to be actively read, producing RNA messages. While some or even most of this transcription might just be the result of all the biochemical hurly-burly going on inside the nucleus, there are more and more examples where it's given a supposedly dead gene a new lease of life. It turns out that the line between dead and alive is much more blurry than we might think. The zombie genes are rising.

❧

One man who has his own scientific zombie story to tell is Howard Chang at Stanford University. I go to see him towards the end of the day and wait patiently outside his office for a seemingly endless line of students to troop out. He's neat and serious, apologising for keeping me waiting but informing me that he has a 'hard stop' at half past four. I waste no time getting down to business, asking him to tell me about his work on pseudogenes.

Like all good horror films, it started innocently enough. Howard and his team were quietly working away in the lab to understand a common medical phenomenon called inflammation. This is the response your body makes to a wound or an infection: cut your finger and you'll soon notice the skin around it becoming red and swollen. Although it may feel uncomfortable, a certain amount of inflammation is a good thing, as it helps attract the attention of the immune system, getting it to come and sort things out. But the exact chain of molecular events that leads to inflammation – and how it goes away again after things have healed – is still unknown.

In search of answers, Howard wanted to find out the repertoire of genes that get turned on and off during inflammation. To find out what happens when it all kicks off, his team grew mouse skin cells in plastic dishes in the lab. Then they lit the fire of inflammation by adding a molecule called TNF-alpha, normally produced in the body in response to damage or infection. Finally, they read all the RNA messages produced by the cells, providing a picture of the underlying pattern of gene activity. But there was a surprise. As well finding plenty of RNAs telling the cells to make particular proteins (as might be expected), they also found a whole bunch that didn't. Unlike the zoo of short fragments that we previously met (see p. 155), these RNAs were much longer. Somewhat unoriginally, scientists refer to them as 'long non-coding RNAs'. 'There are a large number of these, but only a few have been studied,' Howard tells me. 'There are a lot of questions about them, such as what is the full set of long non-coding RNAs? When do they come on? What signals do they respond to, and what might they do?'

Having found them, the team then lined up the letters of the long RNAs they'd found against the DNA in the cells, to reveal where in the genome they'd come from. It was here that they got a further surprise: many of the long non-coding RNAs were coming from pseudogenes. At first the team was suspicious. As the ENCODE debacle has shown (see pp. 23 and 30), RNA can be produced from all sorts of locations in the genome but might not actually be doing anything useful. These were different. 'Not only were the pseudogenes turned on in response to a signal, but the response was exquisitely specific,' he says. 'A certain kind of inflammatory signal, let's say from bacteria, would turn on one pseudogene. Another signal, from a virus perhaps, would turn on another one. The set of long non-coding RNAs from the pseudogenes was so specific that it was an internal reflection of what the cell was seeing on the outside. We could tell what kind of infection the cell was being bombarded with, what kind of inflammatory signals were coming in, by looking at the pattern of these long non-coding transcripts.'

In Howard's mind, the fact that particular transcripts turned up in specific situations strongly suggests that they're important. But it's also pretty strange. Why would these dead genes be coming back to life, and what are they doing? To get some answers, he and his team focused on one pseudogene in particular that had some interesting properties. Rather than playing a role in ramping up inflammation, as most of the other genes that got switched on were, it seemed to be involved in calming things down afterwards.

When you get infected with something, whether it's a bacteria or a virus, special cells in your immune system start pumping out a bunch of different molecules telling cells to start the inflammation response. Some of these work by teaming up with transcription factors, triggering them into action to turn on the genes that get inflammation going, including the pseudogene that Howard's team had found. To their surprise, they discovered that the RNA produced by the pseudogene sticks to one of these crucial transcription factors, effectively tying its hands so it can't do its job. As a result, there is nothing to keep the inflammation genes switched on, so the response dies back down. It's a good example of the kind of negative feedback loop that happens all over the place in biology, stopping inflammation from running out of control or persisting after it's no longer needed.

With more than a touch of whimsy, Howard named the pseudogene Lethe, after the mythical river that meandered through the ancient underworld. 'In Greek mythology it's the river that you pass through to forget about your life. We think this is an analogous situation, and because pseudogenes are essentially dead genes we're evoking that idea,' he explains. 'Let's say the cell encounters some sort of infection and it mounts an inflammatory response. It's gone through this battle and has fought it off – it's time to go back to normal life and forget about the inflammation. In the absence of Lethe, the cell will basically keep going forward with the inflammatory response. It doesn't realise that the battle is over.'

There's another twist to this story. While humans may become more scatterbrained with age, there's a key difference with Lethe: it seems to *lose* its memory-wiping powers over time. Straight away Howard knew this was an important finding. Chronic inflammation that flares up and doesn't die back is a common problem in older age, and has been fingered as a culprit in several illnesses, including cancer and heart disease. 'Prior work we had done showed that as organisms get older, some of the genes involved in inflammation get activated and stay on over time. We didn't know why that was the case. It turns out that the ability to activate Lethe actually decreases with age. Somehow, this negative feedback system is gradually lost when organisms get older. This is a tantalising clue for how, perhaps, the whole system of inflammation and gene regulation in ageing could break down.' An intriguing case of dead genes being needed to help keep us alive for longer, perhaps?

As well as the memory-wiping Lethe, Howard thinks that there are plenty of other pseudogenes that spring back to life to play important roles in our cells. Researchers around the world are finding more and more examples of zombie genes that seem to support this idea. Yet what's not clear is how they work, or even if they all function in the same way. Some might behave like Lethe, producing an RNA that can tie up transcription factors or other important bits of the cell's gene-reading machinery. In other cases, their similarity to the functional gene they're derived from may be important. Because they're copies of normal genes, they have many of the same control switches embedded within them. This means they can compete for the same transcription factors, distracting the attention of the gene-reading machinery away from their protein-coding counterparts. Some of them might act as molecular 'sponges', mopping up micro-RNAs or other small molecules. Then there are others that are still a complete mystery (and probably many that aren't important at all).

What's also not clear is exactly how widespread this kind of control is throughout the genome. Large-scale surveys like ENCODE suggest that a large fraction of our DNA, including hundreds of pseudogenes, can be read into RNA messages, but – as we've seen – this doesn't always imply they're actually doing something. It's going to be tricky to figure this out, because things change in response to the environment. As Howard tells me, 'There's a lot of potential for some of these RNAs to have a role. And I think the point is that it might need a bit of stress, such as inflammation, to trigger some of these events. So our survey may not be complete if we just look at healthy cells during normal development.'

It's now half past four. We've reached Howard's 'hard stop' and it's time for me to leave. He and other scientists are still hunting for more functional pseudogenes, figuring out how they fit into the bigger picture of gene control. But while it looks like zombie genes like Lethe might be helpful to us, there are other undead genetic elements that are more ambiguous in their activities. In any horror movie, the scariest moment is when a corpse starts twitching. And within your genome there are sequences that should be as dead as a doornail, yet they arise and start walking. Or rather, jumping.

CHAPTER SIXTEEN

On the Hop

If you ever go night-trekking in the depths of the forests of northern Brazil, you might be lucky enough to spot a pair of wide-open brown eyes staring back at you from the bushes. They'll belong to a douroucouli or owl monkey – the only nocturnal monkey in the world. As well as looking pretty, these big-eyed beauties have another unique talent: they're the only South American monkeys that can resist HIV infection.

Primates from other parts of the globe – the so-called Old World monkeys that evolved first in Africa and Asia – are resistant to HIV infection thanks to a gene called Trim5. It makes a protein that puts a stranglehold on the virus when it gets into their cells, stopping it from multiplying. Unfortunately, Trim5 didn't survive the evolutionary journey across the pond intact, and it doesn't have the same infection-fighting ability in most species of American monkeys. However, one ancestral douroucouli struck it lucky. At some point in the distant past, a virus-like piece of DNA copied and pasted itself into its Trim5 gene, accidentally bringing along another gene for the ride. The stowaway is a gene called CypA, which can stick to the outer coating of the HIV particles. It turns out that this cut-and-paste hybrid gene is just the right thing to counteract HIV, making the monkeys immune. It's easy to see how this might be an evolutionary advantage, and the genetic interloper quickly spread through the species.

For the douroucoulis, and a handful of other examples dotted around the scientific literature, this genetic jump is obviously and immediately useful. But in most cases it's not, and the new addition is a hindrance rather than a help. It also turns out that genes hopping about is an incredibly common occurrence in genomes throughout the natural world, including within our own.

One of the more surprising observations to come out of the efforts to sequence the human genome is just how much of it is boring and repetitive.* For example, your DNA is peppered with more than half a million copies of a repeated DNA sequence called Line-1, described as 'one of the most ancient and successful inventions in the genome'. The secret of Line-1's success is that it's a transposon – the formal name for a jumping gene. It's worth reading up on the history of the discovery of these remarkable elements by the even more remarkable Barbara McClintock.† Within the few thousand letters of its DNA, Line-1 contains all the instructions it needs to hijack the cell and make three things. Firstly, it tricks the cell's gene-reading machinery into making an RNA message containing a carbon copy of itself. This then encodes the instructions for making some molecular scissors that can snip DNA, plus the recipe for the rule-breaking protein reverse transcriptase which can turn RNA into DNA. This troublesome trio – the Line-1 RNA, the scissors and reverse transcriptase – sneak back into the cell's nucleus, like a gang of biological vandals. Once there, the scissors pick a random place in the genome and start cutting. The Line-1 RNA (which is a single strand of letters) settles itself into the resulting gap, and reverse transcriptase gets to work converting this into double-stranded DNA, seamlessly weaving the invader into the fabric of the genome.

Obviously, randomly snipping up DNA and inserting rogue elements is a risky business. Although much of our genome is tolerable junk or unwanted garbage, which may have a buffering effect against the potential impact of jumping

---

*The genome, that is. Although much of the donkeywork of sequencing and gene analysis is also very boring and repetitive.
†A visionary scientist with ideas far ahead of her time, she was eventually recognised by the Nobel Prize committee at the age of 80 for her work discovering and understanding jumping genes in maize plants. Recognition came so late because her work was rejected by the scientific community for a long time. Partly because it was pretty weird, but partly because of plain old sexism.

genes, there's still a chance that a Line-1 element may hop into an important gene or control region and affect how it works. In rare situations it might be beneficial, perhaps causing a slight change in the organism that helps it to survive and reproduce – like the lucky douroucoulis. In this case it can start spreading through the population on its way to becoming a permanent part of the genome. But, more likely, it'll be useless or positively harmful.

As might be expected, organisms have evolved several ways to stop the hopping. There are the pi-RNAs we met in Chapter 14 (see p. 160), which are particularly important in smaller animals such as worms and fruit flies. Then there's DNA methylation in mammals, which stops embedded transposons from being read into RNA in the first place. Without the initial step of creating Line-1 RNA, there can be no tools for snipping and rewriting it into the genome. However, no defence system is foolproof. There are two times in our lives when our methylation abandons us, leaving our DNA vulnerable to transposon-based vandalism. The first is right at the start of our existence, just after fertilisation, when almost all the DNA coming from Mum and Dad is wiped clean of DNA methylation. This erases any epigenetic baggage that the egg and sperm might have brought with them (apart from a few very important exceptions, as we'll see later), so the embryo can develop normally. Then there's another wave of methylation removal that happens a few days later in the cells that get laid down for the next generation – the germ cells that will one day become eggs and sperm* – which gets rid of any marks that were put on during the earliest stages of development.

Both these waves of demethylation are absolutely vital, but they leave our DNA critically vulnerable to attack by transposons. Although there are other protection mechanisms

---

*I find it staggering to think that the cells that became the egg and sperm that made me were created when both my parents were barely recognisable embryos, not long after my grandparents conceived them. Also a bit icky, but that's reproductive biology for you.

at work to prevent these occasional acts of vandalism turning into a full-scale trashing, you don't get to be the most successful invention in the genome by not taking advantage of your enemy's weak spots. Recent research suggests that up to one in every 20 babies is born with a transposon that has jumped into a completely new place in its genome. As a result, transposons hopping about in the early embryo or germ cells are thought to be an important driver of evolutionary change over time.

Aside from their activity in early development, it's easy to discount these jumping genes as irrelevant to the rest of an organism's life. Our repetitive DNA, including transposons, gets methylated early on in development, rendering it effectively sealed in an epigenetic coffin. But, like the jumpy moment in a horror film when the zombie's eyes snap open, a paper published in the journal *Nature* in 2005 gave the scientific world a start. It came from the lab of Fred Gage – known to his friends and colleagues as Rusty – based at the Salk Institute in California. Rusty and his team were looking at some unusual cells growing in the lab, originally taken from a rat's brain. Known as neural progenitors, these cells can transform into a range of different types of nerve cells. Because of this impressive regenerative capacity, people are very interested in their potential for repairing the damage done by head injuries or neurodegenerative diseases such as Alzheimer's.

To find out more about how they work, Rusty's researchers analysed all the RNA in the nerve cells, revealing a catalogue of genes that were switched on. To their surprise, they found a significant number of Line-1 RNA messages mixed in among all the other transcripts. According to conventional wisdom, the only place these transposons should be active is in the early embryo or in the germline. But, this being Rusty's lab, conventional wisdom didn't apply. The next question was to discover if the transposons were actually on the move, or whether the RNA was just sitting there doing nothing. To find out, the scientists popped a human Line-1 element into the rat cells, carrying along with it a special molecular 'flag' that would reveal if it had hopped from its initial location and landed somewhere else. Sure enough, the labelled Line-1 started jumping. Not only that, but it seemed to show a preference for landing in active nerve genes, even affecting what kind of cells the progenitors would become. And when they genetically engineered mice carrying a flagged human Line-1 in their brain cells, the team saw it spring to life and start moving.

This was big stuff. Although sticking a human Line-1 into mice or rat cells was a relatively artificial system, the results raise the intriguing possibility that transposons can jump around in nerve cells, potentially affecting gene activity and even their ultimate fate. Could jumping genes somehow be rewiring our brains? A tantalising hint that this might be true came when Rusty's lab took a look at human neural progenitor cells. Again, they found that their flagged human Line-1 could start copying and pasting itself around, just like in the mouse and rat brains. Not only that, but when they looked at RNA from human adult brains, they found traces of Line-1 RNA in the hippocampus – the part of the brain used for learning and navigation – but not in similar samples from the heart or liver from the same person.

Although their experiments didn't definitively show that Line-1 elements were actively jumping about in the human brain, they provided a pretty strong hint that they might be. To prove that they really do go on the move, scientists needed to go large. In 2011 a nearly 20-strong international team came together, charged with using the latest techniques to look carefully at the DNA in samples from five regions of the brains of three people. They were searching for places where it looked like Line-1 or other transposons had copied and pasted themselves during each individual's lifetime. Not entirely surprisingly, given Rusty's previous results, they found thousands of places where Line-1 had hopped in. But there was also something entirely unexpected. In almost twice as many places it looked like a different transposon, called an Alu element, had moved in. Alus can't jump by themselves, but they can hijack Line-1's molecular scissors and reverse transcriptase to get themselves on the move. Not only was Line-1 rising up to vandalise the genome, but Alu was getting in on the act too.

🐾

To find out a bit more about what might be going on as the joint starts jumping, Scott Waddell at the University of Oxford is downsizing from humans and mice to look at the tiny brains of fruit flies. We arrange to meet for a chat during the lunch break at a conference at the Royal Society in London. After the morning's talks are over I drag him off to sit on the posh padded leather chairs in the equally posh lobby, refusing to let him get any food until he's answered all my questions.

Despite Scott's disappointment at being deprived of lunch, he's warm and friendly. We chat about Rusty's work on rat, mouse and human brains – 'Rusty Gage! With that name he should be either a porn star or a country singer ... he's got a strong moustache,' he muses. Then he takes me back a few years to when he and his lab were studying the peculiarly named mushroom bodies in the fly brain,[*] where flies make and store their memories. 'We knew that two groups of nerve cells were required at different times to make memories,' he recounts. 'The idea was that they would probably communicate with each other. But one of the great untold secrets is that we didn't actually know what the transmitter systems were in these neurons, so I wanted to try and find them by looking for the genes that were active in those cells.'

Scott and his team looked at all the RNAs being made by the two different types of cells in the mushroom bodies, and noticed extremely high levels of messages from transposons. They weren't Line-1s – those are specific to mammals – but other jumping genes commonly found in flies. He'd always had a passing interest in transposons and had seen the papers on jumping genes from Rusty's lab, so he figured that the flies

---

[*]Because they look like mushrooms, if you squint a bit.

might be doing something similar. 'The most interesting hypothesis I can think of is that this helps generate additional diversity,' he tells me. 'You can generate random differences between cells in that area of the brain, and that might somehow help information processing.'

It's an incredible idea, implying that some – or even many – of the cells in your brain are some kind of genetic pick and mix, each with subtly different characteristics and abilities. But how to prove that it's true? The best way to approach such a challenge would be to sequence all the DNA in single neurons from individual flies, but this isn't easy when you're searching for one tiny element that's jumped out of half a million that haven't. It's a real needle-in-a-haystack job. 'So far it's been done in one study in mammals, so it's doable in principle but it's technically very difficult,' Scott sighs. As a compromise, he and his team are scanning the DNA sequences in small groups of cells from individual fly brains. Right now, all he can tell me is that it looks like the transposons are definitely jumping.

One person who's managed to get it working is Chris Walsh at Harvard Medical School in Boston. He's done a painstaking analysis of the DNA in 300 single brain cells from three different people. His team certainly found evidence that transposons had copied and pasted themselves around the place. It's not very common, though – they see new Line-1 jumps in roughly one in every handful of cells. Although it looks like it might be an uncommon event, the jumping genes seem to prefer to hop into and around important nerve cell genes. This means they're more likely to have an effect, whether positive or negative, than random leaps into the vast swathes of junk within the genome.

Added together, all these results point towards one thing: in some brain cells, at some point, in several species, the normal controls that keep transposons switched off are released. Rusty thinks that this is tied into the time when new nerve cells are born in the brain, adding an extra random element that subtly alters the capabilities of each cell. Like the effects of transposons throughout evolution, some of these changes will be useful and make the brain cell function better, some will make no difference, and some will leave the cell worse off than its neighbours. All of it contributes to the uniqueness of each person's brain. But there might also be other, more sinister, impacts.

'The other thing that's interesting is that we found higher transposon activity levels in older adult brains,' says Scott, just before I release him to attack the conference buffet. 'Other people have showed that transposon levels continue to go up with age. So I think an attractive idea is that the jumping that we see in adults could keep going in older fly brains. Over time you would pepper the genome with bits of rogue genes, and eventually the flies will have problems with brain function.' Although it's just a theory at the moment, it's possible that jumping genes may play a key role in the way our brains tend to run down over time in unpredictable ways.

And there's more. Over 21 million people around the world are afflicted by schizophrenia. It's a distressing mental health condition that tends to strike teens and twenty-somethings in their prime, profoundly changing their thoughts and dramatically altering their chances of living a normal life. An identical twin has a 50 per cent chance of developing schizophrenia if their sibling has it, telling us that there must be a significant genetic component, but also a hefty chunk of nurture, too. So far, the hunt for genes involved in schizophrenia has yielded less than impressive trophies: a handful of the usual suspects that tend to turn up in every search for psychiatric disease genes, and precious little else. Then at the beginning of 2014 a team of Japanese scientists discovered higher numbers of Line-1 elements in the DNA of brain cells from people with schizophrenia, particularly around some of the genes involved in wiring the connections between brain cells. The results aren't a definitive smoking gun, but they do suggest that gene-jumping might play a part in creating the neurological chaos of schizophrenia.

It certainly looks like transposons are on the hop in the brain, either as part of normal development or when things go awry. It's an open question as to where else in the body these jumping genes may be active. Because it requires poring over the DNA of single cells to spot them, nobody's really been looking. And could they be involved in other diseases where genes are disrupted, such as cancer – a disease that starts with individual cells picking up faults in their genes and going rogue? This idea isn't new. There's a fascinating paper from 1992 by scientists in Japan and the US, who were searching for faults in a gene called APC in patients with bowel cancer. Normally, APC helps to protect us against cancer, so inheriting a faulty version of it brings a significantly increased chance of developing bowel tumours. But the disease can also be caused by randomly occurring faults in the gene that crop up during a person's lifetime. In the hope of spotting some interesting-looking APC faults, the researchers looked at the DNA sequence of the gene in 150 of these random, non-hereditary bowel tumours. To their surprise they found something nobody had ever seen before: in one of them a Line-1 element had hopped right into the gene. Presumably, this caused so much damage that APC could no longer work properly, leading to the unlucky patient's cancer.

Those researchers made their discovery the hard way, using old-fashioned sequencing techniques to slowly piece together the story. Now times have changed and cancer scientists are busily reading every genome they can get their hands on. The International Cancer Genome Consortium and other groups are sequencing DNA from tumours and healthy tissue from thousands of cancer patients around the world. And because each person's cancer is a microcosm of evolution, starting from one or just a handful of cells, if a Line-1 or other transposon has hopped into an important cancer-related gene and affected how it works, then these approaches might reveal it. One recent paper looking at oesophageal cancer genomes showed that there are around 100 Line-1 jumps in an average tumour sample, with some having as many as 700 new insertions. Maybe jumping genes are a potent cause of the genetic chaos in cancer, maybe they're not. It will be interesting to see what the sequencing reveals.

There are also intriguing hints that a bunch of jumping genes known as endogenous retroviruses are going on the hop very early on in development, when we're just a handful of tiny cells. Although they make up around eight per cent of our genome, it's not clear yet exactly what they're up to (or whether they're good or bad for us), so it's definitely a field to watch for the future.

Finally, allowing transposons to rise up and move around the genome seems like a risky business. If people like Scott and Rusty are right, then activating jumping genes to increase the differences between nerve cells is worth the trade-off in a potentially increased chance of cancer, neurological disease or just going dotty with age faster. It's worth pointing out that any genetic changes in adult cells don't get passed on to the next generation – only the germ cells, the egg and sperm, have that privilege.* What happens in the brain stays in the brain, to steal a slogan from Las Vegas. Problems like cancer and dementia tend to mainly affect older people, and evolution doesn't really care what happens to you when you're old. It only cares that you are smart enough to make and raise babies. So perhaps passing on the capacity for unleashing jumping genes in order to make a more interesting brain that helps you learn, grow and get laid is more beneficial in evolutionary terms than any potential risks later in life.

Only time – and lots of fiddly single-cell gene sequencing – will reveal the answers. Regardless, the discovery that our genome might not be a perfectly fixed template in every cell highlights an important point that many people don't realise about biology: it's wobbly. *Really* wobbly.

---

*OR DO THEY?? We'll find out later.

CHAPTER SEVENTEEN

Opening a Can of Wobbly Worms

The fault, dear Brutus, is not in our stars but in ourselves.
William Shakespeare, *Julius Caesar*

If you wanted to pick a nice simple organism to study in the lab, you could do worse than the tiny nematode worm *Caenorhabditis elegans,*[*] known as *C. elegans* for short. Checking in at just 1mm (1/25in) long and made up of a precise 1,031 cells for an adult male, these animals have been the subject of intense scientific scrutiny since the late 1960s. Curiously, every single worm is made in exactly the same way. From the minute an egg is fertilised, the cells of every growing baby worm divide and specialise in an identical fashion. It's so stereotyped that each individual cell in a worm even has its own name.[†] Researchers can watch worm embryos growing and developing under the microscope, seeing new cells pop into existence right on cue, and it was the first animal to have its genome completely sequenced.

If you wanted to pick a nice place to investigate these little wrigglers, you could do worse than the Centre for Genomic Regulation in Barcelona, perched on the eastern Spanish coast. I visit on a clear January day, walking along the seafront from the subway station with the soupy Mediterranean rippling endlessly beside me. The sky is baby blue over the grey sea and I have to hold my hands up to shade my eyes from the weak but persistent winter sunshine. I'm here to talk

---

[*] Latin for 'nice rod', apparently.
[†] Sadly, these are boring names like AB, CP1 and DA7 rather than Albert, Carrie and Doris.

to Ben Lehner, a researcher who's found a peculiar glitch in the biological clockwork of these supposedly perfectly predictable worms. He's young – boyish, even – dressed in jeans and a hoodie with a mop of brown hair and a casual demeanour. The kind of guy I'd expect to see tapping away on a shiny laptop in a Dalston coffee bar rather than hanging out in a paper-strewn office opposite a bustling lab, telling me his thoughts about how our genes work.

'We're all different,' he says, poking a finger towards his chest, then at me. Apart from a shared taste in sneakers and the basic head, two arms, two legs stuff, he and I have little else obvious in common. 'But where do these differences come from?' The obvious answer that springs to mind is that it must be in our genes. 'So how different is my DNA to yours?' I ask him. 'Four million differences, roughly, plus a chromosome,' he shoots back. 'Biologists tend to think of all vertebrates as pretty much the same. We use mice as a model for humans, because for most of the important properties we're the same. We have the same body plan, we have the same development. Humans are very similar to each other, but then you have this huge variation on top of it, and that's the interesting question.'

The accepted explanation for this variation is the combination of nature plus nurture – our genes plus our environment. A good example that we're all familiar with is identical twins, nature's own cloning experiment. They should (in theory at least) have exactly the same genes. But any parent of twins will tell you that their offspring are not perfect carbon copies. At this point, people usually invoke the magic word 'epigenetics', pointing to the influence of the environment on our genes causing differences in gene activity. Although twins tend to grow up in very similar environments, they will obviously have plenty of unique experiences along the way, which might be enough to explain the differences between them. Case closed? Not according to Ben.

'You have the genetic differences that you inherit from your parents, but then every time a cell divides there are new mistakes that happen. So you will have some stuff in your

DNA which neither of your parents had – a very small number of changes that happen each generation that are unique to you.' To paraphrase Philip Larkin, not only do your parents fuck you up by filling you with the flaws they had, they also add some extra mutations, just for you.* Some of these happen when eggs and sperm are made, or arise very early on in development. Then there's the stuff that happens later in life: chance mistakes by the machinery that copies your DNA when cells divide, random lightning-strikes by chemicals or radiation, and accidents that happen due to the general hurly-burly of the chemical reactions of life. It's this genetic misery that leads to cancer if it happens to knock out important genes. And less serious mutations can have more subtle impacts on our health and physical appearance. 'Each of the cells in your body is very slightly different in terms of its DNA sequence. Cancer is primarily caused by these mutations,' he explains. 'They're the ultimate chance events – completely random, and you can't predict them.'

To understand this biological roulette wheel a bit better, Ben's been looking at the ultimate clones – *C. elegans* worms. Yet, as he's discovered, there's quite a bit more going on than just occasional DNA faults. 'When we make a mutation in a worm gene, sometimes the outcome seems random. If you have 100 genetically identical individual worms all carrying the same gene fault, 90 of them will be affected in some way but 10 won't. They'll look normal. Then for a different mutation, say half of them will be affected and half of them won't.' All the worms have come from the same parents, so their genes should be identical. It doesn't seem to be down to random mutations – he's checked. And it can't be due to the environment, because they're all living in the same Petri dish. It's not nature. It's not nurture. So it must be something else. 'Magic?' I ask. He laughs and scuffs his trainers on the floor. 'Not magic. There must be a chance event that's happening at some level.'

---

*Scientists prefer to use the less sweary term *de novo* mutations.

To figure it out, he's going right back to the start of worm development – the beginning of the journey from one cell to many. A multitude of genes are switched on and off in response to all sorts of chemical signals and feedback loops. Signals triggered when the sperm enters the egg tell gene A to make protein A, which switches on gene B, which makes protein B, which switches on gene C and turns off gene A and so on and so on, creating a dazzling network of biological circuits. When a cell ends up with the right combination of signals at the right time, it makes a decision to do something – to become a nerve cell or a skin cell, say – or to divide again and wait for a different fate. Repeat this a thousand times and you've made a worm. Repeat it millions and millions of times and you've made a human baby. In this way, life creates order out of the chaos of biochemistry.

'Worms are kinda freaky,' Ben tells me, shaking his head in bewilderment. 'Every cell always behaves in the same way. Every cell has a name. We know how it divides, we know what it will become. They're super-invariant. But still when you make these mutations it becomes variable.' Ben and his team have been taking a closer look at their wobbly worms, using genetic engineering techniques to remove specific genes one at a time. 'What we've shown is that when you have your favourite disease-causing mutation, the degree to which other genes are being switched on and off can actually be very important for whether that mutation has an effect or not. When you mutate one gene, the system is partly broken. So then these small differences in the levels of gene activity start having stronger effects.'

Biology is not binary. Switching genes on and off isn't like flicking a light switch up or down, flipping from pitch black to blazing light. It's more like a dimmer dial, moving gradually between completely off and full beam, through every shade from a romantic evening glow to daytime brightness. This makes sense if you think about how genes are read into RNA. For example, a gene could be a little bit active, only making one RNA message per minute. Or it could make two. Or ten. Or a hundred. Or it could be really going for it and churning out thousands. Exactly how many depends on a whole range of things, from the number of transcription factors that are available at the time to a hefty dose of random chance. Remember Bickmore's baubles from Chapter 9 (see p. 97)? That's the idea that transcription is the result of enough stuff being in the right place at the right time, rather than a precisely controlled interaction. There's a big difference between making one RNA message and making a thousand, but what about 10 versus 20? Or 15? If the gene is encoding a chemical signal that's important in development – such as the difference between being a skin cell or being a brain cell – then a relatively small tweak in the amount of gene activity could make a massive difference to the fate of the cell.

At this level, Ben explains, genetics is a numbers game. 'The gene we've studied the most is a transcription factor, encoding a protein that switches on other genes.' I ask for more details, notebook at the ready. 'Doesn't matter what it's called,' he says. 'The only reason we chose it is because the outcome is fifty-fifty and it makes it easy to study. When we make worms without it, half of them die and half of them are absolutely fine. It's a chance event.'

The next step was to carefully measure the activity levels of a whole range of different genes early on in development in these wobbly worms, to see if chance fluctuations were related to whether each one survived or not. What Ben and his team found was striking: if the activity levels of two particular genes in the worm embryo are high – meaning that they are making loads of RNA messages – then nine times out of 10 it will survive to adulthood. But if the levels are low, with only a few messages being produced, then the chances are that the worm isn't going to make it. 'So we've gone from a fifty-fifty outcome to being able to say 90 per cent are going to be able to survive, just by measuring two bits of the system,' he tells me, holding out his palms and shifting them up and down like a set of weighing scales. 'Still not 100 per cent, though,' I point out. 'No. I think there are likely to be other components of the system that we need to find, but these are difficult experiments to do. We're trying to quantify stuff super-accurately but having to use techniques that are a bit rubbish.'

I admire his honesty. 'So what's going on?' 'That's a great question, and that's what we're trying to understand.' He gives me an example from the world of bacteria, where this kind of cell-to-cell variation in gene activity has been studied more closely. Although they may all seem identical, even individual bugs are slightly different from each other. 'When genes are turned on they do little bursts of transcription. You turn the gene on, you make a few RNAs, you turn it off. It's kind of – tsch! Tsch!' He flashes his fingers at me like firework bursts. This gets even more haphazard in complex organisms like worms or humans, where enormous amounts of biological machinery – RNA polymerase and all its associated molecular bells and whistles – have to be assembled in order to start reading a gene. The polymerase runs along the gene, making RNA for a bit, and then it stops and falls off. To make another transcript, the whole lot has to be built again. This is where biology becomes chemistry and rubs itself right up against the laws of physics, and the capacity for 'quantum jitters' – as news reports describe the results of a 2015 study revealing the wobbliness of the structures of DNA and RNA at the atomic level – is significant.

In the case of proteins, some are stickier than others. Interactions can also be encouraged by packing lots of the right molecules into the same place in the cell, but it's still essentially a matter of chance. As I've noted before, scientists prefer to use the term 'stochastic', as it sounds a bit more impressive than 'random' when admitting the truth: we're only here because of haphazard molecular shuffling. 'It's a crowded intracellular environment,' Ben explains. 'It's extremely complicated. There are molecules bombing into each other and moving around, and that's the ultimate cause of all the stochastic things we see. It's just a question of how far back we have to go.'

In the case of animal development, it's all the way back to a single fertilised egg. This biological singularity – one cell, one genome – is the most exciting and dangerous time in any organism's life, as tiny random differences in gene activity here will be amplified as the embryo grows, echoing throughout its whole life. Some of this might be down to tiny genetic variations picked up when the egg and sperm are made. It could also be due to varying levels of proteins and RNA that come ready-packed in the egg and sperm, or just random wobbles in gene activity as things get started. So why risk leaving something as vital as this to chance?

'C. elegans embryos are fantastic to study because they go from one cell to more than 500 in a few hours, and the error rate is tiny,' he explains. 'Every cell divides in the right way, in the right place, moves to the right place, makes the right connections. And they can do it under all kinds of different conditions. You can change the temperature and it still works. You can squash it between glass slides' – he squishes his palms together by way of demonstration – 'and it still develops fine. Biological systems are incredibly robust.' He touches his fingertips lightly together and looks intently at me from under a scruff of hair. 'Everything is connected. The really big question is why does biology not go wrong?'

Ben thinks that the random wobbliness in gene activity levels gives some of his worms a chance to overcome the effects of the bigger genetic alterations that he's inflicted upon them. Worms (and every other living thing) have to be able to withstand at least a certain amount of change to their environment. We can think of these alterations as external 'mutations'. The mechanisms that give the worms the ability to withstand these external changes probably help protect them from internal genetic mutations, too.

This philosophy provides some insight into one of the problems that plagued me and my colleagues back in the lab, as well as many other researchers working in genetics. We'd spend weeks/months/years genetically engineering mice with missing or altered genes or control switches. Then we would wait for the pups to be born with great anticipation, coming in early in the morning and at weekends to check on the precious occupants of our murine maternity ward. Sometimes there would be no mouse babies born at all.

This wasn't great news, but it suggested that whatever DNA sequence we'd knocked out did something really important earlier in development, which could then be investigated further. In the best-case scenario, we'd get live animals that would have something obviously wrong with them, mimicking humans with the same genetic problem. Quick, write a paper! The most frustrating outcome of all was a cage full of mice that appeared to be totally normal. An obvious conclusion was that the region of DNA that we'd removed did nothing, as far as anyone could tell. But these were lab mice, living in warm, well-fed luxury. It became a standing joke between me and my colleague Rob that whenever we heard about a genetic alteration that apparently had no effect, we would turn to each other and ask sagely, 'Ah, but did they take their mice to the moon?'

While bagging their rodents a place on a moon mission isn't possible for anyone but the most well-connected scientists, it's a silly way of making an important point. The way that geneticists find out how genes work is by breaking them, but a lot of researchers never bother to find out if their altered animals are still normal under more challenging conditions. Imagine using this approach to understand how a motor car works. You can cut the fuel line and see an immediate dramatic effect: the car won't go at all. Pull off the steering wheel or cut the brake cable and the vehicle will start moving just fine, but you'll soon run into obvious problems. Then you notice a small wire under the dashboard and cut it. Nothing happens. The car runs just fine. Ah well, you shrug, it probably wasn't important. Then you head out of the garage and into the garden to enjoy a glass of Pimm's in the summer sunshine. But once the chill of winter arrives and the windscreen heater doesn't work, you realise what you've broken. Although if you live on the sunny California coast, where the thermostat is permanently set to pleasant, you might never ever notice.

A good example of this principle in action comes from a gene going by the strange name of Shavenbaby, responsible for directing the growth of tiny hairs on the back of a fruit fly maggot. Without it, the little larvae are naked and hairless, hence the name. Next to the Shavenbaby gene lies a whole bank of control switches that make sure it gets turned on in the right place at the right time to create the maggot's hairdo. In the spirit of 'break it and see what happens', US researcher David Stern and his team tried removing these switches one by one. The answer was: nothing much. Each of the switches seemed to do pretty much the same thing, telling the gene to come on in the same or overlapping groups of cells on the maggot's back.

If breaking one thing at a time doesn't do anything, the geneticist's solution is to break some more. For his next experiment, Stern took out three switches at once. Again, much to his surprise, nothing happened. But then he took his flies to the moon – or more accurately, to the most extreme temperatures they might expect to encounter in the outside world, rather than the cosy room temperature of his lab. Finally, *something happened*. At a fresh 17° or a toasty 32° Celsius, things started to fall apart. The maggots missing their switches were noticeably balder, although they still managed to grow a few wisps of hair. Fly larvae with a full complement of control elements had a luxuriously hairy back, regardless of the temperature.

The story of Shavenbaby highlights the robustness of biological systems and their ability to survive environmental and genetic changes. Messing about with the activity levels of different genes or changing the temperature might be fine most of the time, as there are all kinds of backups and compensation mechanisms that keep things running. There may be many slightly different duplicate versions of a gene encoding very similar proteins, which can substitute for each other. But if there's too much random wobbling, or the backups are also compromised and can't compensate, then things fall apart. Ben's experiments show that you can break genes a little bit and others will step in to compensate. But break the genome more and more, and the cracks start to show.

'We make one particular mutation in our worms and 90 per cent of them are OK. Then we make a mutation in a different gene in some other worms and 90 per cent of them are OK too. But if we combine the two faults together we end up with 100 per cent death. None of them survive.' So what does he think is going on? 'At the end of the day it's all acting through the same molecules,' Ben says. 'If you change something in the environment, you're changing the signals that come into the organism. If you make a mutation, you're just changing it from the inside. Any mechanism that is going to make you robust to the changing environment is also going to make you robust to your changing genetics. And there's a pretty powerful evolutionary pressure to keep it.'

Luckily for us and everything else on the planet, life doesn't work like a simple electronic circuit. Although there are some genes that are absolutely essential, there are many that aren't. You can't just break one part and automatically expect the whole circuit to fail. The fact that biology is shaped by evolution, rather than design, makes it able to cope with a lot of fuzziness around the edges. And, frankly, we need the wobbliness of biology just to get through the tricky business of being alive. 'There are errors in everything,' Ben points out. 'Errors in copying DNA, transcription and splicing; mistakes in translation; problems with protein folding; errors in protein modifications. Biology has to be able to deal with quite high percentages of stuff being wrong. Molecular biologists tend to think of things in engineering terms, that it's a very precise system.'

He warms to his theme. He tells me about an article he read describing how geneticists might study a radio to figure out how it works – *i.e.* by breaking stuff and seeing what happens. 'The guy who wrote it said that genetics is stupid because you can never properly study how a radio works by just pulling bits of it apart – if you pull out the power switch on a radio you break the whole thing.' But this isn't what Ben sees when he looks at his worms, and in his view it misses the point entirely. Living organisms are not consumer goods. Someone sat down and designed a radio, incorporating a sensible and straightforward electronic circuit that can be easily broken, rather than the multiple redundant genes in most living systems. This makes sense from a business perspective: why build endless fail-safes into a radio when you don't need them? That costs money, and it's not efficient. Regardless of the protestations of believers in intelligent design – a position so intellectually contorted it belongs in a philosophical circus – nobody designed biology, and life does not run like a slick machine. Yet it still works remarkably well. 'It was made by small mutations and big mutations. It was made by genes evolving over time, and it's incredibly robust.' And in Ben's view, as long as we look carefully – and combine enough genetic mistakes in the right way – we can start to understand how our biological circuits were patched together by evolution.

By now I've talked with Ben for an hour and a half. Before I leave to spend all my money in the Spanish designer outlet in Barcelona airport, I return to the point where we started: what makes us all different? 'Fundamentally, the reason why mutations have different effects in different individuals is the genetic background,' he explains. 'In any two human genomes there are millions of differences. And of course the effect of any one of those can depend on any of the others, on how they interact together. Human geneticists and statisticians hate this idea, because it really complicates everything. They like to think of every mutation having one effect, which is independent of everything else.' One mutation, one effect. One gene fault, one disease. It would be nice and neat, wouldn't it? Yet everything we know about biology says that it's completely wrong.

CHAPTER EIGHTEEN

Everyone's a Little Bit Mutant

Gholson Lyon is one of those big, slightly brash Americans. When I meet him at a conference, I guess straight away that he's a medical doctor. Rimless glasses, grey suit jacket, beige chinos, brown shoes. After too many years living and working in hospitals I can spot them a mile off. He's a psychiatrist by training, but has since turned to hunting for genes involved in brain disorders – such as Tourette's, autism and schizophrenia – from his lab at Cold Spring Harbor in New York. He has a reputation as a provocateur, and as I listen to his presentation it's refreshing to hear someone speak their mind in an academic environment where opinions are often carefully couched in caveats.* As the first speaker after lunch, he starts with a punch to shake us out of the food coma and put us in our place, pointing out that recent celebrations of the 60th anniversary of the discovery of the structure of DNA by Watson and Crick might have been overly triumphalist. 'The Earth is 4.5 billion years old. The dinosaurs went extinct in an asteroid impact on the Yucatán peninsula 65 million years ago. Of course, we humans like to think that we've come a long way in the last 150 years, yet our understanding is so minuscule to what it will be going forward. As long as another asteroid doesn't come and destroy us.'

When he was starting out on his medical career, Lyon met a lot of kids with Down's syndrome. This is a genetic disorder caused by having an entire extra copy or substantial bonus chunk of chromosome 21, effectively providing three copies of the genes on that chromosome. Most of us will be familiar with the rounded face and distinctive features of children with the condition. But there are health problems, too.

---

*The intellectual gloves tend to come off in the bar afterwards.

Learning disabilities. Sight defects. Heart disease. Gut problems. Infertility. The list goes on and on. But it's not a precise checklist: every person with Down's has their own mix of health issues, affecting them to different extents.

It was this variation that struck Lyon. How could these children, who all have the same genetic problem, show such different symptoms? It was a question that continued to bother him as he moved on to work with kids with a different genetic disease: DiGeorge syndrome, also known as velocardiofacial syndrome, caused by a missing portion of chromosome 22. Again he saw a huge range of different outcomes, from severe intellectual disability and facial deformity to mild effects, all due to the same gene fault. And the more closely he looked, the more variation he saw.

One day, while he was working in Utah, a family came to him with a distressing story. Since the 1970s, many of their baby boys had been born with a distinctively strange appearance. Each had thick eyelids, large ears and eyes, and a beaky nose. Their skin was loose, wrinkling around a tiny body, encasing a weak heart that stuttered and skipped as it beat. Tragically, every single affected child died by the age of two. Straight away Gholson realised he was dealing with a new genetic syndrome. 'We called it Ogden syndrome,' he tells us, scanning round the faces in the warm, slightly-too-small lecture hall, 'because the family lives in Ogden, Utah. I got into a lot of arguments with the other researchers in the team because they each wanted to name it after themselves.' There's a ripple of self-conscious recognition from the geneticists in the room. 'Seriously, the whole point is to honour the family and to say it's ancestry-dependent. This mutation in this family from this town causes this disease, and the name itself is embedding information about the syndrome. Whereas to call it Lyon syndrome does absolutely nothing to impart any information.'

Forsaking their egos, Lyon and his colleagues set about tracking down the faulty gene responsible for the problems. Eventually they pinned it on a single DNA letter change in a gene on the X chromosome in every affected baby boy.

None of the girls born in the family showed any signs of the syndrome. This is because females have two X chromosomes. The normal X acts as a genetic backup, although they can still pass the mutation on to any male children. Boys have just one X and a Y, so there's nothing to compensate for the mistake.

Like the kids with Down's and DiGeorge syndrome, Lyon noticed that although all the Ogden babies were very similar, they weren't exactly the same. They had subtle differences in their faces, and death came to claim them on a variable timetable – one baby at just a month old, another making it to two years, the rest somewhere in between. This was puzzling, given that they all had exactly the same gene fault and were all part of the same family. What Lyon was seeing was a more complex version of Ben Lehner's fifty-fifty worms (see p. 187). In that situation, random wobbles in the levels of gene activity and chance variations in the worm's genes were the difference between life and death. But because humans are a lot more genetically diverse than highly inbred lab animals – and we live in a far less controlled world than a worm in a plastic dish – there are a lot more potential wobbles in our genome and environment, hence a wider spectrum of effects. And this is where things start to get very complicated indeed.

For years genetics researchers have been looking for genes involved in human diseases by studying DNA from afflicted patients. Along the way they've found a whole bunch of genetic suspects, unlocking our understanding of many diseases caused by mutations in single genes, such as cystic fibrosis and Huntington's disease. They've tracked down gene faults that massively increase the risk of certain types of cancer, such as mistakes in the BRCA1 and 2 genes that are linked to hereditary breast, ovarian and prostate cancers. But, just as Lyon noticed with the DiGeorge and Ogden kids, there's still a big variation in how the effects of these faulty genes play out in any given person.

The problem is that we know practically nothing about what's in our genes on an individual basis. Sure, we have 'the human genome', but this is an idealised genetic Bible for our species, cobbled together from a handful of people. The DNA in your body differs from that script in millions of ways, as does mine. And it follows that looking for DNA faults and variations for a particular disease solely in people who are affected by it gives a wildly misleading picture of what's in our genome and how it works, whether in sickness or in health.

To prove the point, Gholson Lyon tells us about a paper that came out just two days before in the prominent journal *Nature Genetics*. A large team of scientists in the Netherlands read the entire genomes of around 250 Dutch family trios – Mum, Dad and a child – just to see what was in there. What they found was surprising, to say the least. As Lyon succinctly puts it, 'Holy crap! Each person has a whole bunch of gene faults that are thought to cause rare diseases, although they're perfectly healthy. So we know *nothing* about most mutations in humans.' For example, two of the people in the study had major mistakes in both copies of a gene that normally makes a protein which helps to protect the structure of the lungs. Without it, the delicate tubes and air sacs start to break down, causing serious breathing problems. The effects of the faulty gene normally kick in at around 30 to 40 years of age, but both people in the Dutch study had made it well into their sixties with no lung problems at all. As the researchers themselves put it in their paper, 'These results highlight the potential pitfalls ... and the challenge of interpreting personal genomes.'

There may be a glimmer of hope, though. In January 2015 researchers at Imperial College London published a paper looking at people with a condition known as dilated cardiomyopathy, in which the heart becomes enlarged and doesn't work so well. In some cases it runs in families, causing apparently healthy people to suddenly drop down dead. This is the kind of story that makes front page news when it happens to a strapping footballer, and many people with the condition never know they have it until the worst happens. Scientists have long been searching for gene faults and variations linked to dilated cardiomyopathy, but although quite a lot of candidates have been unearthed, each of them only accounts for a handful of patients. Then in 2012 there was a breakthrough. A collaborative effort by US and UK researchers revealed that about one in four cases of the condition could be explained by a fault in a gene called Titin – a molecular spring that gives heart muscles their bounciness. And what a gene it is! A giant of the genetic world, Titin encodes the recipe to make the largest known protein produced in the human body. But this epic size was a problem, making it difficult to study its DNA sequence in large numbers of people cheaply using old-style technology. This changed with the advent of next-generation sequencing, so the team set about reading the Titin gene in lots of people. Then things started to get confusing.

Having been elated to find that a quarter of dilated cardiomyopathy cases might be explained by a faulty Titin gene, the researchers were concerned to find that around one in 50 completely healthy people in the general population also seemed to have a similar mistake in their Titin gene, which looked like it should be harmful. Were they about to drop dead of heart failure, or was there something else going on? To figure it out, Imperial researcher James Ware and his colleagues looked at more than 5,000 people, from a range of backgrounds. Some had severe dilated cardiomyopathy and were waiting for a heart transplant, others had been referred to a specialist clinic after having heart trouble, and others were (as far as anyone knew) completely healthy. After reading the entire DNA sequence of the Titin gene from all of them, the team mapped the genetic variations and faults back to the state of their health. Eventually a more sophisticated picture emerged.

There were some versions of the gene that didn't seem to be associated with heart disease at all, even though they might look bad on paper. That's good news for anyone who turns out to have one of these 'lucky' Titin gene faults, as they won't need endless trips to the doctor to keep an eye on the condition of their ticker. But there were others that were strongly linked to the chances of developing dilated cardiomyopathy. Bad news for the people in the study who already had the condition, but good news for their family members, who can now be tested and monitored for any problems – a potentially life-saving early warning.

Stories like this highlight one of the major challenges of genome research: predictions are only as good as the data they're based on. Previous research may suggest that certain gene faults are harmful, and this information gets duly entered in the big online databases that gather data about this kind of stuff. But just like the lucky Titin mutations, which should be harmful but actually aren't, it may be wrong. It's only by looking at large numbers of people, both sick and well, and carefully mapping that information back to their health that true associations can be teased out. Sometimes the effect of a particular gene fault might be cancelled out or enhanced by other variations in the genome. Or it might be influenced by random wobbles in our biochemistry, like the fifty-fifty worms.

To put it bluntly, everyone's a little bit mutant. We're all walking around with a handful of potentially dangerous or even deadly gene faults, as well as millions of random little differences peppered throughout our genome. For some people these problems will always reveal themselves – like the tragic Ogden boys – yet there will still be a spectrum of how severely someone is affected. The thing that makes the difference is what else is in our genome, which determines how well we can withstand life's assaults on our DNA. Depending on what other cards you've been dealt in the unique genetic hand you got when your Mum and Dad made you – plus any faults you picked up along the way, and the effects of the environment you live in – you might never even know. Until, that is, someone starts ferreting around in your genome.

This revelation seems to have come as a shock to many people, particularly those who are fixated on the neat, easily explained idea of one gene fault always leading to one disease. This kind of thinking has come as a result of the huge advances made in molecular biology, which have given us a highly compartmentalised, reductionist view of how genes work. In this world-view, a gene is a recipe that tells the cell to make a protein involved in some biological process. If the protein is missing or faulty, then the process won't work and the person will get sick.

There are examples of diseases where this seems to be the case, such as cystic fibrosis, albeit with a greater or lesser amount of variation between patients. But, as Ben's wobbly worms and the data from the Dutch family study and others like it show, this isn't the whole story. It shouldn't be a surprise, according to Gholson Lyon, if only people would *read*. 'If you do not know history, then you are doomed to repeat it. And if you do not read, then you are hopeless,' he rails at us. The historians in the audience nod smugly, the rest of us shift guiltily in our seats. 'People are only just beginning to realise what the early geneticists have said for over a century – that any given mutation can only be figured out in the context of the genome and the environment and the entire organism. Ancestry matters. Your genetic background matters. The reason that I called it Ogden syndrome is that it implicitly says ancestry matters. You can't just take a mutation, pluck it out of one family, pop into some other family and expect to have any kind of correlation between genes and their effects.'

There's an even bigger issue when it comes to conditions that can't be pinned on a single faulty gene. You can't have missed the headlines over the past decade claiming that scientists have found tens or even hundreds of 'new genes' involved in heart disease, obesity, autism, schizophrenia, cancer and more. In fact, what these studies usually turn up are tiny variations in DNA, known as snips,[*] between people affected by a disease and their healthy counterparts. But each variation only accounts for a tiny fraction of the risk of getting a condition. For example, people who have a letter T at a particular place in their DNA might be very slightly more likely to develop schizophrenia than those with a C. And

[*]More correctly written as SNPs, short for Single Nucleotide Polymorphisms. These are single DNA letter differences between two people at specific points in the genome.

there are more than a hundred similar variations, each with a similar tiny increase in risk.

Millions of pounds – hundreds of millions – have been ploughed into these kinds of experiments, known as genome-wide association studies. But despite trawling through bigger and bigger populations, pulling out more and more variations with ever smaller effects, the results have told us relatively little that's practically useful. Most of the DNA changes, about four in every five, aren't even in protein-coding genes. The assumption is that they're in the control switches that regulate when and where genes are turned on and how active they are. Lots of snips linked to schizophrenia and autism are in or near genes that are turned on in the brain. No surprises there. But that doesn't tell us anything about what these genes really do, until people start doing the hard work of investigating them in detail in the lab.

The other big problem with this approach, particularly when hunting for genes involved in psychiatric and neurological conditions, is that it's incredibly difficult to precisely define each disease. One child's autism is not exactly like another's. Depression is a big umbrella used by psychiatrists to cover many manifestations of misery. Schizophrenics don't all show exactly the same symptoms. It's a lazy assumption that each disease is caused by the same processes going awry in all cases, according to Gholson Lyon. 'I am happy to report that I definitely discovered that these are incredibly complex and extremely heterogeneous diseases,' he says, somewhat sarcastically, 'to the point of the word "depression" being utterly meaningless, the word "schizophrenia" being utterly meaningless. Anyone who tells me they're doing a study on depression or schizophrenia or autism without any sort of meaningful sub-typing, I stop listening at that point because they're actually working on something that does not exist.'

The latest buzzword in genetics is 'stratification', carefully selecting ever more tightly defined patient groups in order to boost the chances of finding stronger associations between genes and diseases. This might mean only picking women with depression whose condition has similar characteristics and starts at roughly the same time in life. But while this might yield a few more interesting links, I'm increasingly feeling that there are limited further gains to be made. At a time when mental health services in the UK are being cut to the bone, ploughing the cash into developing better care might ultimately make more of a difference than finding a snip that increases someone's risk by a few per cent. To be fair to the snip-hunters, their discoveries do sometimes provide a useful chisel for researchers to start prising open the biological processes that underpin a disease. Not many people want to do that, though, because it's hard. It involves doing tricky experiments, often using animal models, and taking years to carefully unpick what's going on. Much easier to apply for a million-pound grant and go fishing for yet more snips instead.*

According to Gholson Lyon and others who think like him, the only way we'll start to truly untangle the complex relationship between what's in our genome and how it affects the way we eventually turn out is by reading whole genomes. Lots of them, from anyone we can get our hands on, and then comparing the information in the DNA to what its host looks and behaves like. This approach can be roughly summarised as 'Sequence All the Things'. There's the 1,000 Genomes project in the US, which has done what it says on the tin and then some. Not to be beaten by the Americans, there's also the 100,000 Genomes project kicking off in the UK to sequence DNA from cancer patients and children affected by rare diseases. Realising that there's more to humanity than Western white folk, scientists are now reading genomes from people of diverse ethnic backgrounds all over the world, and China is getting into DNA sequencing in a big way. As might be expected, Lyon is a big fan of this large-scale approach. But there are two big issues at stake. One is technological, the other cultural.

*I'll get off my soap-box now.

When I was first working in a lab, round about the turn of the millennium, reading even small stretches of DNA took ages. We would inch our way across regions of the genome we were interested in, getting four or five hundred letters every few days and pasting them together to reveal the complete code. Despite having been jazzed up with automated machines and clever computers, DNA sequencing technology was until relatively recently still based on Fred Sanger's original 1977 method, and there was a limit to how fast it could go. Even the massive sequencers at the Sanger Centre outside Cambridge – cranking out reams of human, animal, plant and bacterial genomes in stark rooms humming with the song of their enormous air conditioners – just weren't quick or cheap enough to cope with our lust for letters.

Then came the next generation. Faster, cheaper and more accurate, these machines smash DNA into tiny fragments and read them all at once, using computer programs to assemble a meaningful sequence from the resulting chaos. But although the costs are coming down and the speed is going up all the time, many researchers are choosing to only sequence the bits of the genome that are actively read into RNA (known as the exome). Although this gives a good overview of the quirks and variations within an individual person's genes, it may miss crucial changes in the control switches in between the genes, which regulate them. Given that most of the genetic variations linked to human diseases are in these control regions, it seems foolish to ignore them.

Now we have a further leap in technology. Next-next generation, I guess you could call it. The company Oxford Nanopore made headlines in 2012 when it announced a DNA sequencer roughly the size of a mobile phone, which pulls individual strings of DNA through a tiny electrified hole rather like pulling a string of beads through your fingers. Different DNA letters disrupt the current in slightly different ways, giving a readout of the sequence as it goes through. At this rate, it's soon going to be realistic to think about reading entire genomes from not just thousands but maybe millions of individuals.

And it's here that we come up against the biggest issue. Large-scale genome sequencing is not a technical problem. It's a people problem. 'Our entire field was done a great disservice by the people who hyped up the human genome project,' Gholson Lyon complains. 'They said, "We'll sequence the genome and then within a decade we'll solve all these illnesses." The fact of the matter is that because of that, there's absolutely zero political will in America to give more money to genome sequencing.' Instead, the US government is investing huge sums of money into mapping all the connections within the entire human brain – 'an infinitely more complex task' – and praises us Brits for our government-funded 100,000 Genomes project.

'I believe that technically we're only five years away from getting a highly accurate whole genome,' he preaches. 'I hope that in 20 years time every newborn – at least in the industrialised world – will be getting their genome sequenced. But I don't know if we're socially and culturally even 20 years away.' There are big issues here that people feel very strongly about. Some are only too happy to have their genomes sequenced. I spoke to a representative of an organisation for families with children affected by rare genetic diseases, who told me that most of them are jumping at the chance to get some kind of insight into their child's condition. Maybe they're hoping for an explanation for their kid's symptoms, or the name of a recognised syndrome to write on a request for extra support at home or school. At best there's an outside shot at finding a cure, perhaps by finding a fault in a biological pathway that can be targeted or tweaked with an existing drug.

There are also the recreational genetics fans. These are the kind of people who spit in a tube and send it off to a company like Google-backed 23andMe for analysis. The results can tell you exactly what percentage Neanderthal you are, figure out where in the world your ancestors came from and even help you to trace your family tree, as well as offering information about the genetic risk of a whole range of diseases, from

Alzheimer's to cancer. In December they were given free rein to flog their wares in the UK, although their legal situation in the US is more restrictive for the time being. Given the web giant's love of big data, it seems inconceivable to me that this might not provide a rich seam of information for genetics researchers to mine. And given that the fascination with genealogy seems to set in around the menopause, based on a sample size of one (my mother), perhaps the future of genome sequencing will be driven by middle-aged women?

Yet for every person who is happy to turn their DNA over to a company on the internet or to the UK government, there are others who are deeply concerned. There are big questions that aren't easy to answer: Who owns your genetic data? What will they do with it? Who can see it? And who can buy it? This isn't strictly a personal matter either, because your genes aren't just yours. You share them with your blood relatives. If you have a gene fault that significantly increases your risk of bowel cancer; then your parents, siblings or children might have it too. Do they want to know, or should they just be told anyway? I don't know the answers to these questions, and my opinions about them change from time to time. But I think it's vitally important that we all understand more about what our genes can and can't tell us, in order to make decisions that affect our lives, health and families.

The idea of 'a gene for' a disease or characteristic is hopelessly simplistic, and there's a huge amount we don't know about how the information encoded in our DNA (our genotype) gets translated into what we look like, how we behave and what diseases we get (our phenotype). Right now, there's a huge black box sitting between the two. It's time to prise it open.

CHAPTER NINETEEN

Opening the Black Box

When the human genome sequence was first published more than a decade ago, many scientists believed – and frequently told the public – that once we knew all the letters of our DNA the secrets of life would suddenly become clear, like finding the solution to a particularly tricky cryptic crossword. But being able to read the book only tells us a limited amount about how we're made. Just as a cake recipe in a cookbook can't control the precise temperature of your erratic oven, know the exact size of your eggs or guarantee the taste and texture of the crumbs on your tongue, there's many a slip 'twixt the genetic recipe in our cells – our genotype – and the living, breathing reality of our phenotype.

To understand some of this complexity we need to take a step back to the old days, when it all seemed so simple. What we understand today as the field of genetics properly kicked off in 1865, when a monk called Gregor Mendel published a paper entitled 'Experiments on Plant Hybridisation' in an obscure German journal. Scientifically minded from a young age, Mendel had initially wanted to study inheritance by breeding mice. However, this was felt to be too sexy (literally) in the eyes of the local bishop, so he ended up pottering about with some peas in the gardens of St Thomas's Abbey in Brno, now in the Czech Republic.

Pea plants were a good target for Mendel's curiosity as they could be fertilised by hand, with the help of a paintbrush, so he knew exactly which plants had 'done it'. By carefully crossing together plants with distinct characteristics – such as white or violet flowers – Mendel discovered that traits seemed to be inherited in a specific way. Breeding a purple-flowered plant with a white one created purely purple progeny, leading to the assumption that whatever was responsible for purpleness was somehow stronger than whiteness, and dominated over it. But crossing these child plants incestuously together led to something odd. Rather than a sea of violet-hued pea plants, for every three purple offspring there was one with white flowers. Some sort of instruction for white flowers had sneaked across a generation, but could only emerge under certain circumstances.

Mendel figured that there had to be two versions of the flower-colour recipe: one white and one purple. He also realised that these are passed independently from parent to junior pea. White-flowered plants must have two copies of the white version – one from each parent – while pure-bred violet-flowered plants have two purple ones. When two different-coloured plants are crossed together, their offspring inherit one recipe of each type, but purple wins out so they all have purple flowers. Mendel referred to this as 'dominance', with the weaker white characteristic termed 'recessive'. But what's going on in the next generation, when white makes a comeback?

The eureka moment came when Mendel realised that the colour instructions must be staying separate in the offspring rather than being blended together, contrary to popular theories of heredity at the time. Crossing these hybrid plants together gives four possible mixtures of instructions: purple and purple; white and white; and two versions of purple and white (one where purple comes from the male, the other where it comes from the female). And because purple is dominant over white, three of these plants will have violet flowers while only one will bear white blooms (the one with two whites).

Unfortunately for Mendel, not to mention the wider scientific community grappling with the nature of the gene, the journal in which he published his findings, *Proceedings of the Natural History Society of Brünn*, wasn't exactly a big hitter in the world of scientific periodicals. As his master work vanished into obscurity, Mendel turned his attention briefly towards breeding killer bees (allegedly)* before being promoted to abbot and getting distracted by fighting with the local government about taxes. It was only at the turn of the 20th century that his work was rediscovered and popularised, notably by the impressively moustachioed William Bateson, laying the foundations for our current understanding of how genes work.

The beauty of Mendel's idea was its simplicity. You get one set of traits from Mum, and one set from Dad, and they get randomly mixed up and passed on to the next generation yet stay distinct. Furthermore, they segregate in defined patterns and exert their influence in an organism according to whichever trait is more dominant. His brilliant deductions defined the rules of inheritance that still hold true today for characteristics and conditions caused by a single faulty gene. For this reason, students usually start here when they first learn about genetics. I certainly remember learning about Mendel and his peas in my own biology classes at school, untidily scrawling capital and lower-case letters on the page in vibrant purple ink† to denote dominant or recessive versions of genes. My classmates and I scrutinised each others' ear lobes to see if they were fully attached to our heads or hanging free, fed each other bits of paper impregnated with chemicals that only some of us could taste, and mischievously tried to figure out if any of us might be illegitimate by looking at our families' blood groups – all classic Mendelian traits that are seemingly inherited in clearly defined ways in accordance with his laws.

It's important to remember that Mendel and the other early geneticists had no idea of the physical nature of the information they were tracking down the generations. All they could see

---

*Some sources claim that Mendel bred bees that were so vicious they had to be killed. While it's true that Mendel did keep bees at the abbey and was frustrated by the difficulties of getting them to mate according to his scientific whims, there's no good evidence to support this story. Bah.

†No academic reason, I was just one of those annoying show-off kids who liked to write in purple.

was the outcome in the organisms they studied, whether it was peas, fruit flies or anything else. They knew that *something* had to be responsible for shuttling traits from parent to offspring, but had no idea what it might be. That didn't stop people coming up with all kinds of ideas of what the physical vehicle of heredity might be, from sensible things like proteins (a good guess, but wrong) to the bizarre concept of the homunculus – a miniaturised human packed inside every sperm. Eventually, thanks to an enormous amount of dedicated research over the 20th century, we arrived at the concept of the molecular gene as many people understand it today: a string of DNA encoding the instructions to make a specific protein or RNA. Each gene is a distinct entity, which fits with Mendel's idea that traits don't blend. He only saw purely purple or white flowers, not a range of hues. But although one gene might make one protein, it's not as simple as saying one gene makes one physical characteristic. The journey from gene to physical trait is a tortuous one, with lots of wobbliness and complex interactions along the way.

The problem is that humans are not peas. And even Mendel's peas aren't really like peas. While the scientific world was going crazy for Mendel's ideas in the early 20th century, English biologist Raphael Weldon was unconvinced. For a start, he was deeply suspicious of Mendel's pea-plant data and how neatly they fitted the theory, bitchily musing in a letter to a colleague that 'I can't help wondering if the results are too good?' On top of this, Weldon felt that Mendel's conclusions weren't a good reflection of the true nature of peas (or genetics more widely, for that matter) because he only went for the simple stuff. Flower colours may seem to segregate beautifully in a Mendelian manner, but most other characteristics don't.

To make his point, Weldon published a photograph of a neatly arranged crop of dried, podded peas generated by crossing yellow-seeded pea plants with deep green ones. According to Mendel's ideas, each pea should have either been green or yellow. If you squint at the picture and feel kindly towards Mendel, you could almost agree that each orb was purely one colour or the other. Weldon wasn't prepared to do that. Instead, he saw a whole spectrum of colour ranging from butter yellow through to grassy green. Fifty shades of peas, indeed.*

It's not just peas. You only have to look around your own family to realise that you and your siblings, if you have them, aren't a perfectly proportioned split of characteristics from Mum and Dad. My two sisters and I are not peas in the proverbial pod, but a mishmash of family traits blended together, with plenty of apparently unique features thrown in for good measure. Heredity is complex and mysterious, and subtle variations in genes and their control switches (as well as the random hand of chance) all work together to influence how we turn out.

At this point it might be tempting to throw your hands up in the air, declare that genetics is all just too complicated and go straight to the pub. Instead, I went to see one of the researchers who's trying to figure it out.

❧

Stephen Montgomery is tall, dark and (if my ears do not deceive me) Canadian. In his lab at Stanford University, he and his team are unpicking how variations between human genomes translate into variations between people themselves, prising open the black box between genotype and phenotype. It's a challenge he feels rather acutely, having recently received the analysis of his own genome. It turns out that he's got more Neanderthal DNA in him than 99 per cent of the rest of us – something that's hard to believe by looking at his sharply chiselled cheekbones and neatly cropped hair. 'It changed my whole view on what Neanderthals were like!' He declares to me. 'Everyone throughout my whole life has been describing Neanderthals as savage brutes, and I think now they're just

---

*The rediscovery of Weldon's peas owes much to the persistence of science historian Greg Radick at the University of Leeds.

really misunderstood.' He laughs defensively. 'We're very gentle! It's just the victors that write the history books ...' 'We humans clubbed them to death, stole their women?' I suggest. 'Yeah, you're horrible people!'

Troglodyte ancestry aside, Stephen enthusiastically explains to me what he's trying to do. Using the latest DNA sequencing technology and massive computers, he's trying to work out how tiny variations in the vast expanse of our genome affect when and where genes get turned on. This is something that until very recently geneticists could only dream of. We're not talking about the big things – major alterations that completely break a gene or its control switches – but the subtle differences that dial gene activity up or down just a little bit. Those small changes can make all the difference.

As Stephen tells me, it's not just about figuring out why I have a slightly snubby nose while that of my younger sister is pointier. Seemingly minor variations in our DNA code can have a big impact on our health and well-being. Like the mythical butterfly flapping its wings and causing a hurricane, a few rogue letters in the genome might eventually result in the chaos of disease. 'For instance, someone might be able to fight off an infection slightly better than someone else. Or someone responds to a drug better than someone else,' he tells me. 'We want to be able to describe those benefits so that they might be able to help us understand what makes each of us unique, but also to understand how we might use that to tailor any particular treatments for any individual.'

To date, a lot of the effort in working this out has focused on the small fraction of the genome making up the protein-coding genes. Not only is it a much smaller bucket of genetic material to be rifling through, but it's also relatively easier to understand how a tweak to a finished protein might affect its function. All proteins are made according to the same rules we learned right at the start of this book – each three letters of DNA encoding one particular amino acid – making it possible to predict the impact of any changes in the underlying genetic sequence. For example, altering one single letter in the middle of a protein-coding gene might inadvertently throw a 'stop' sign into the middle of the resulting RNA message. Then when that message is translated to make a protein, the resulting product is truncated and stubby. Alternatively, changes that result in a switch from one particular amino acid to another within a protein can alter its physical properties and ability to function properly. Although it can take a lot of fiddling about in the lab to figure out the impact of any given DNA alteration on the resulting protein, it's by no means impossible.

All well and good, but around eight out of ten genetic variations linked to disease don't lie inside the protein-coding bits of genes. They're in the control switches, the junk, the rubbish and even the garbage. There's a whole heap of stuff here we simply don't have a clue about, as Stephen tells me. 'For the non-coding genome, we don't really know much about its architecture. We know that there are transcription factors that bind to the control switches. There are non-coding RNAs, and there's all sorts of gene regulation. We're trying to piece this together.'

The way he and his team are doing this is by going large. They're collecting big groups of individuals and looking at two things: firstly reading the entire genome sequence from every person, and then analysing the varying levels of activity of all their genes by carefully measuring the amount of RNA produced by each of the 20,000 or so genes in various tissues. But it's not quite as simple as just taking a single DNA sample – from the blood, for example – and getting all the answers. It's very easy to get hold of blood, which is why it's so commonly used for DNA analysis.[*] But if we only look at blood we're not going to be able to understand all the disease-related genetic variation in the rest of the body. 'Some genes are highly active and some are not very active at all, and we now have technologies

---

[*]Although there's no DNA in mature human red blood cells, there are plenty of white blood cells prepared to give up their genetic cargo.

where we can actually profile this all across the genome. But it's very dependent on what tissues you're looking at,' he explains, with a slight air of frustration. 'So if you're looking at pancreatic tissue, you might find very high activity of insulin and related genes. And if you're looking at blood cells you might find a lot of RNA for haemoglobin. To understand the impact of genetic variation on the body, you also need to account for the context of the tissue that you're looking at.'

To try to unravel some of this complex genetic tapestry, Stephen is profiling many different tissue types in lots of groups of individuals, looking for correlations between specific little variations in the DNA sequence, and figuring out if these map on to small changes in gene activity levels. Because this has to be done by looking for statistical associations between the two – spotting whether DNA variation X is more likely to be associated with RNA level variation Y – across millions of letters of DNA, thousands of genes and hundreds of people, it's a phenomenal exercise in number-crunching. 'The change in DNA sequencing technology – I know it's clichéd – but the capabilities that it has given us are just enormous! It used to be big genome centres that had this sequencing stuff, but now individual labs have the ability to generate huge amounts of data. We've got a sequencer through there' – he points at the office wall towards the lab beyond – 'and it's just incredible the amount of data we can get off it. In a day we can read 80 billion bases.' 'How on earth do you sift through all of that?' I ask, incredulous. 'How big are your servers?' 'Not big enough!' he sighs. 'We can generate data so much faster than we can analyse it. I can spend a couple of days in the lab and have enough data to keep us busy for a year.'

I tell him about some recent efforts I've seen to outsource this kind of work to the non-labcoated general public through citizen science projects. One of them was a mobile phone game where thousands of players flew spaceships along paths plotted through real DNA sequencing data flowing from a lab at the Cancer Research UK Cambridge Institute. It was phenomenally successful, taking a month to crunch through patterns that would have taken the scientists six months to trawl on their own. 'That sounds cool! Maybe if we could hook our data up to Candy Crush we could solve our entire problem ...' He trails off, clearly plotting.

Although sifting through the data to find these associations is arduous, it's not particularly hard once the right computer programs are on the case. What *is* more tricky, however, is proving that a particular spot in the DNA is genuinely having an influence on the level of activity of a nearby gene. Recalling the story of the Hemingway cats, there's no guarantee that a control switch will be conveniently located right next to the gene it affects. It could be miles away, in genomic terms. Or the level of gene activity could be influenced by something like a micro-RNA, produced from a gene on an entirely different chromosome. How on earth does he figure that out, I ask? 'We're starting to think about all these pieces that are part of the genome, herding all of these different elements together. That should help us more precisely understand the consequences of different DNA variations on gene activity. Right now it's easy for us to understand the effect of a single letter change. The question is, how do we start integrating richer information?' How indeed.

One approach is to add together all the possible information he and his team can gather about what's going on in the genome: DNA variations that look like they're affecting the docking sites for transcription factors or other relevant proteins, chromatin modifications such as DNA methylation or histone tags, non-coding RNA and more. By mashing together all this data and searching for links that seem to consistently jump out, he and his team can slowly start to tease out the threads that join the information encoded within our DNA to the real-life end product.

As if that wasn't complicated enough, in addition to looking at different parts of the body in different people there's also an added layer of complexity: geography. Each population in the world has its own mix of genetic variations, shuffling through their genomes at different rates according to the whims and vagaries of human migration and mating.

While some of these are responsible for differences in physical characteristics or disease risk, others might just be an unrelated DNA change that happens to be common in the population – an irrelevant junk mutation in the junk DNA. To get round these problems, Stephen and his team are casting their net widely across the world, carefully analysing DNA from people of different origins to separate the genetic variations that genuinely affect gene activity levels from the random regional variations that don't. And it doesn't stop there. Not content with looking at variations in genes in hundreds of tissues from thousands of people all over the world, researchers are now taking genetics into the fourth dimension: time.

Although it's often tritely said that all our cells have the same DNA,* this isn't strictly true. The genome in that single fertilised egg cell marking the start point of your existence is very slightly different from the DNA in a randomly selected skin cell in your hand holding this page or a brain cell pondering these words. Individual cells pick up unique typos in their DNA as we go through life, caused by damage from internal and external sources, slip-ups in the copying process as cells divide or maybe even rogue jumping genes going on the hop. And then there's general epigenetic wobbliness: proteins stick onto or fall off control switches in the genome, DNA methylation patterns shift and change, histone modifications come and go, messenger RNAs are spliced and edited in different ways, non-coding RNAs are made and destroyed …

Yet from this seeming biological chaos our bodies create order – on average, the output from all this randomness is remarkably stable. You keep on living, breathing, thinking, eating and doing everything else that you need to do. Sometimes things go wrong, of course. A classic example is cancer, caused by the accumulation of mistakes affecting the

*Including within the pages of this book.

genes that control when cells divide or die. But more broadly, there's an ever shifting panoply of genetic and epigenetic changes going on throughout the body as we age and (if we're unlucky) succumb to disease.

Just as you don't have the same photo in your passport when you're 40, 60 or 80 as you did when you were 20, we shouldn't think of analysing someone's genome as a one-off snapshot. Stephen's colleague at Stanford, Mike Snyder, is offering himself up for regular genetic testing, looking at changes in the levels of activity of thousands of genes as well as any alterations in the underlying DNA. Mike calls this project the 'Snyderome', reflecting the idea of his own personal genome profile over time, although on my travels I've heard people jokingly refer to it as the 'narcissome'. They're probably just jealous. For his part, Stephen is less harsh. 'We know that there are a lot of changes as people age and it's different for men and women, from childhood to puberty to menopause. There's potentially a lot of value in being able to track people over time, but that's also very challenging to do because people have to be committed to it. With Mike being a geneticist he's deeply committed to these ideas, so he collects data on himself to show that these kinds of changes can be observed and whether that might lead to something meaningful.'

Where things start to get really interesting is when we bring technology into the mix. Along with several of my friends, I now wear an unobtrusive silicone wristband that tracks my activity and reports back to my mobile phone, which does the equivalent of tutting disapprovingly at me when I'm being lazy, eating crap or drinking too much wine. Apple has produced its own souped-up version of this gadget: a smartwatch that can monitor your heart rate as well as constantly bugging you with your emails. Developers are busy working on wearable technology to track blood pressure, blood sugar levels and all sorts of other parameters.

This so-called 'measured self' or bio-monitoring movement may be giving geeks everywhere a whole new set of graphs to play with, but it's also a geneticist's wet dream. 'We've never had this kind of data,' Stephen tells me, his eyes shining. 'We don't know what the genetic effects are on this data over time. Even when you go to the doctor you're just getting a readout of this one snapshot in time ...' 'Like when I go to the doctor my blood pressure is enormous, because I get stressed about going there?' I suggest. 'Exactly! There's all these things that influence it. Right now you might be very relaxed because you're not around all the chaos of your normal life, so your blood pressure is really low, but the rest of the time it might be a lot higher.' I laugh hollowly. Luxuriating in genetic chit-chat in his sun-soaked California office seems a million miles away from my usual domestic drama. 'It's all these types of things that potentially we might be able to do with genomic technology in the future. So as well as being able to survey your blood pressure and so on over time, you may get a very detailed understanding of what the behaviour of your genes is at any point in time and that will inform you to a much higher degree about your overall health and well-being.'

Combining the power of modern genetic analysis with bio-monitoring could be a potentially life-saving way to spot the earliest signs of illnesses such as diabetes, heart disease and cancer. But I do have a question. Like Michael Douglas in the film *Wall Street*, standing on a beach shouting into his ridiculously oversized mobile phone, bio-monitoring devices and smartwatches are currently toys for the few, rather than tools for the many. 'So you're saying the next wave of genomics is going to come from the fitband-wearing nerds?' 'I think Apple or whoever needs to send these devices all over the world – I think there's a lot of opportunity!' he replies.

But, as Stephen tells me, although we're getting really good at DNA sequencing, we still have very little idea about how variations in our genes affect how our bodies are built or how they work, either in sickness or in health. As he sees it, the key to solving this problem is more data. And this is where the fitbands come in. 'Collecting this information over time as people are just living their lives, and doing that in a way that people feel comfortable with, allows us to understand and to predict powerfully. It would be much better if there were devices that gave people warnings of things to look for, instead of being too late when you're already in the hospital with a heart attack.'

The other big question is whether all this extra knowledge will make a difference. We already know a lot of what the bad stuff is in our lifestyles, but people still smoke tobacco, drink too much booze and swerve the salad bar. He laughs and shrugs with an 'eh, what can you do?' kind of expression. 'I often get asked this by family members who don't really understand the stuff that we're working on. They say, "Oh, have you cured cancer yet?" And my response to that is always yes, of course we've cured it, you just need to eat healthier, stop smoking and exercise more. But that's not the answer they're looking for!'

For now, and probably for many years to come, Stephen and others like him are patiently unpicking the relationship between the recipes in our genes and the non-stop biological bake-off that makes a living body. But as well as this four-dimensional analysis – DNA, gene activity, geography and time – there's another thing that hasn't really been mentioned yet, which adds yet another layer of complexity. Remember right at the beginning of this book I explained that all your cells contain two copies of every gene (apart from that business with the X and Y chromosomes)? And that in most situations we could ignore this and assume they both act the same? Well, for at least a few of them, this isn't the case. All genes are not created equal.

CHAPTER TWENTY

Blame the Parents

A decade ago I was lurching towards the end of my doctorate at Cambridge University. It was several years of thankless grind in the lab, punctuated by fleeting moments of elation when an experiment actually worked, set against a backdrop of poverty and vicious hangovers. The only thing that sustained me was the scientific mystery we were trying to solve. Put simply, it's this: why can't you have a human virgin birth? It may sound like the kind of mischievous question designed largely to annoy religious fundamentalists, but there are plenty of creatures in the world that manage to do just fine without a male in their life. Some lady lizards produce lizard babies without all that messy sex business – a phenomenon known as parthenogenesis – as do lots of insects and some fish. So why don't we?

One man who became obsessed with this question is my old boss, Azim Surani. He's a quiet, thoughtful man who looks every inch the Cambridge professor – bespectacled, bearded and with a ruffle of fluffy grey hair. He welcomes me back into his office at the Gurdon Institute with open arms. Pleasingly, I note that the quality of coffee in the lab has improved vastly over the intervening years.

'I think a lot of science is driven by obsessions people develop for various reasons, and they don't know why,' he tells me, settling back in his chair and steepling his fingers. 'I was working on something completely different, but for some reason I got very excited about this.' As a young graduate student in the early 1970s Azim started working in the lab of Bob Edwards, who would go on to win the 2010 Nobel Prize for his part in the development of *in vitro* fertilisation. He describes Edwards as a rather hands-off supervisor, wrapped up in his work with fellow IVF pioneer Patrick Steptoe. Like a bored child who goes astray while their parents' attention is diverted elsewhere, Azim fell in with Matt Kaufmann – a doctor-turned-researcher at Cambridge who was fascinated by parthenogenesis – and caught the virgin birth bug. There must have been something in the air. At the time, the rise of second-wave feminism was leading some people to suggest that perhaps we could do without men altogether. 'Yes,' he says, grinning at the memory, 'there was a little bit of that as well. There was a lot of debate that females can do everything and that males are just parasites and don't do anything. So that became my obsession.'

To understand why we might struggle to build a woman-only feminist utopia, we need to look at a bit of basic developmental biology. All of us start off with the meeting of two special cells, egg from Mum and sperm from Dad, with each carrying half a set of DNA. There's pretty much all the same genes in each set, but there will be variations between them. Once the egg is fertilised, these two halves have to be unpacked and organised together to make one single genome. Animals that can reproduce parthenogenetically are able to make babies from two sets of Mum's DNA alone, but (historical myths aside) we know of no cases where virgin birth has successfully happened in humans. In rare cases, something can go wrong during the process of egg-making in women, leading to an 'empty' egg lacking any DNA at all. If this gets fertilised, it will now only contain DNA from Dad. The resulting tumour-like mess of cells is known as a hydatidiform mole. Conversely, mouse eggs that have been chemically tricked into starting development without a sperm don't even make it halfway through pregnancy. With only DNA from Mum to go on, the embryos die because they can't form a proper placenta.

Back in the early 1970s there was a lot of debate about how the curious business of parthenogenesis might work, and why it *didn't* work in mammals. Some people thought that the male sperm brought something vital to the mix, either in its DNA or in some other form. Others thought that adding together DNA from Mum and Dad was essential to counterbalance any dodgy or broken genes on each side. Using two sets of Mum's DNA means that there's no way of compensating for this double dose, so the baby won't develop properly.

To get some answers, Azim and his incredibly skilled technician Sheila Barton decided to carry out some unusual experiments with very early mouse embryos, at the stage where they're just a microscopic ball of cells. 'In the early experiments we were doing all sorts of crazy things. We thought that if you combine lots of different parthenogenetic embryos together at a very early stage of development you could make a sort of chimera, and that they would compensate for each other. Sheila and I would mix up these embryos with the idea that if they were born they would arrive on Christmas Day.' Faithfully, he turned up at the lab several Christmases in a row to see if there was a baby mouse born of a virgin waiting for him. Alas, the scientific prophecy turned out to be inaccurate in this case, so by the mid-1980s they had switched to an alternative approach: mixing and matching male and female genomes.

First, they needed to get hold of some fertilised mouse eggs. It's fiddly, but similar in a way to IVF in humans, mice will produce eggs in response to hormone injections. These can be carefully flushed out of their fallopian tubes after mating. If you were to look at them under the microscope, you'd see beautiful transparent spheres floating in space. And inside, you'd be able to just make out two smaller, slightly grainy spheres. These are the pronuclei – two half sets of DNA, one from each parent. Then comes the fiddly part. Using a tiny glass pipette connected to a long rubber tube, controlled by mouth using the gentlest of sucks and puffs, Sheila carefully switched the female pronucleus of one egg with the male from another, and vice versa. This created eggs that had two sets of Mum's DNA (referred to as parthenogenetic) or two sets from Dad (androgenetic).

These were all carefully transplanted back into the womb of a foster-mother mouse, to see if nature would take its normal course. Amazingly, despite all the poking and prodding, the reconstructed embryos got on with the job of trying to make a baby mouse. But around halfway through the foster-mothers' pregnancies, something went wrong. 'I still remember, it was a Saturday morning with Sheila and I looking at these andros and parthenotes. And what immediately jumped out was that they had these opposite phenotypes.' They saw that all the embryos made from two sets of female DNA looked fine by halfway through pregnancy, pretty much as expected for that time in development (although definitely a bit on the small side). But they had virtually no placenta. On the flip side, the androgenetic embryos looked terrible – little more than a poorly formed splodge of tissue, although their placentas looked great. As might be expected from these problems, neither the androgenetic nor the parthenogenetic creations resulted in living pups.

'I realised that one way of explaining this was that there would be genes which would be marked differently in the male and female,' Azim recounts – the implication being that these marks would somehow tell the genes to be switched on if they came from Mum and off if they came from Dad, or vice versa. 'So I wrote this paper and sent it to the journal *Nature*, using the word "imprinting" to describe the phenomenon.' At the same time Davor Solter, a Croatian-born scientist at the University of Pennsylvania, had done similar experiments and come to the same conclusion, so the race was on to be the first to publish. But Azim's choice of words was holding things up. 'One of the reviewers took great objection to us using this term imprinting. And his or her point was that imprinting is used in the context of animal behaviour[*] – we would have to take the word out.' He shakes his head sadly,

---

[*]Developmental psychologists use this term to describe the way that baby animals pick up behaviour from their parents at an early stage in life. This explains why ducklings can convince themselves that a person or even a Wellington boot is their mother, and why workers at a panda conservation centre in China dress up in giant panda costumes while working with the cubs. Actually, that might just be because it makes for awesomely cute photos on the internet.

recalling the frustration. 'I was still trying to set up my lab and I was thinking, well, maybe it doesn't matter. But I thought about it more and realised that it *was* the right word, just a new context. It really captured the concept, so then I just stuck my neck out. I said to the editor that the word has to stay, otherwise I can't proceed with the publication.' Recognising that they were sitting on a significant discovery, the editor agreed. Azim's paper was published, with Davor's following in a rival journal a couple of months later. A scientific field was born: genomic imprinting.

As is often the case with exciting new ideas in science, although the data from the mix-and-match embryos was solid, a lot of people didn't believe it. Azim smiles wryly as he remembers the reception he got, even on home turf. 'I was asked to give a seminar in Cambridge, because they had heard that there was this crazy guy doing strange experiments out on the Huntingdon Road. They were hearing all these rumours so they thought, "We'd better call him." They were very, very sceptical and said this is all phenomenology and it's just weird.'

It was a fair point. At this stage nobody knew what these imprinted genes were, or how it all worked. And it was certainly weird. But he was absolutely sure it was true because he and Sheila had seen it so many times, and someone else had done it, too. 'It was fortunate that Davor's paper also came out. Then people started saying, "It must be real because it's confirmed by another lab and it's not just one crazy guy."' However bizarre it seemed, imprinting looked like it was true. And the next obvious question is, how does it work?

Azim suspected that each of the two copies of certain genes must be marked in some way, depending on whether it had come from Mum in the egg or had arrived in Dad's sperm. These marks must presumably be put onto DNA when eggs and sperm are made. However, these markings must also be erasable, so they can be wiped off and reset according to the sex of the resulting baby: females need to remove any male-specific signals from their DNA, so they make eggs with solely Mum-type marks on both copies of each imprinted gene, and vice versa for boys. But although it seemed feasible that this was the basic principle by which imprinting could work, nobody knew what these marks might be – or, indeed, the identities of any imprinted genes. So the hunt was on to find them.

As with so many things in the history of science, the identification of the first imprinted genes happened by accident. In the 1980s, many researchers were starting to explore the brave new world of genetic modification, creating 'designer mice' carrying extra genes. To make them, DNA containing these bonus genes (known as transgenes) is carefully injected into a fertilised mouse egg, where it randomly shoves itself into the genome and starts doing whatever it's meant to do. Thanks to Sheila's skilful hands at manipulating delicate early embryos, Azim's lab was perfectly placed to start making transgenic mice.

A new student in the lab, Wolf Reik, was keen to get in on the action. He was interested in studying certain families of transgenic mice, hoping to understand more about the control switches that could turn the extra genes on and off. Each family carried an antibiotic resistance transgene known as CAT, originally borrowed from bacteria, which had hopped into their genome in a different place. As might be expected when you start shoving bacterial DNA randomly into a mouse's genome, some odd things were happening. But the oddest of all were the mice from a family called CAT17.

When Wolf looked at baby mice that had inherited the transgene from their Dad, everything seemed fine. The gene was switched on, making the antibiotic resistance protein just fine. But if the transgene came from Mum, then nothing. Nada. Zip. Even more weirdly, when these mice grew up and had pups of their own, the activity (or lack of it) could switch round. If a female mouse had inherited the transgene from her Dad, it was switched on fine in her body but shut down in her young. The opposite thing happened the other way round: male mice who had got a switched-off transgene from Mum were able to somehow reactivate it in their sperm, making babies that all had the gene turned on. This observation made no sense to most geneticists. Why would genes be turned off if they came from Mum, but switched on if they came from Dad? Of course, it was immediately clear to Azim what this pattern meant: the transgene must be sitting in or near an imprinted gene. 'It was a completely serendipitous discovery!' he laughs. 'We found one transgene that did exactly what you would expect of an imprinted gene. It was switched off when it came from the female, and on when it came through the male. And it was reversible, which is what it had to be.'

But just observing these unusual animals wasn't enough. Azim and Wolf wanted to discover the molecular secret of imprinting: they needed to find a physical marker that could be put on and taken off again as genes passed through the germline, one that correlated with whether the gene came from Mum or Dad and was switched on or off. That mark turned out to be methylation, which we met in Chapter 10 – that nubby little methyl group stuck onto the letter C of DNA. Indeed, when Wolf looked closer, he discovered that the CAT17 transgene picked up DNA methylation if it came from Mum but not if it came from Dad. Importantly, these methylation marks got wiped off and reset in eggs and sperm, just as would be expected for imprinted genes. More of these oddities were soon found, including imprinted transgenes that worked the other way round – Dad's version switched off, Mum's version turned on – with appropriate methylation patterns to match.[*] This suggested a clear way in which imprinting worked. When germ cells – the precursors of eggs and sperm – are being made, any existing DNA methylation marks at imprinted genes get wiped off. Then new marks are put on, specific to whichever sex the cells find themselves in. And so the cycle continues from generation to generation.

[*]The observation that DNA methylation correlates with silenced imprinted genes has been the source of much subsequent epigenetic mythology, as it turns out not to be true across the wider genome – see Chapter 10.

While Wolf's peculiar transgene certainly seemed to be behaving like an imprinted gene, it wasn't the same as finding the real thing. That changed when a talented geneticist called Anne Ferguson-Smith joined the lab in the early 1990s. She teamed up with Bruce Cattanach, a researcher at Harvard University, who had managed to get hold of some rather strange mice. Due to a genetic quirk, the animals ended up with two particular bits of their chromosome 7 either from Mum or from Dad. For normal genes this switch-up shouldn't matter – as long as you've got two copies of each, from whatever source, you're cool – but in this case it did. Mice with a double dose from Dad grew into fat little things, much larger than they should have been, while animals with two regions from Mum were abnormally small. Clearly, whichever genes were lurking in that stretch of chromosome 7, at least some of them had to be imprinted.

Anne put her skills to good use and homed in on an interesting-looking gene in the region called Igf2, which encodes a protein that tells cells to grow and multiply. It soon became clear that the gene was switched off if it came from Mum and turned on if it came from Dad, hence the opposing effects on the mice. A double dose of Dad's Igf2 gene meant a double dose of this growth factor, leading to big mice. But if both copies came from Mum then neither was switched on, leading to unusually tiny pups.

More discoveries came thick and fast, with clusters of imprinted genes turning up all over the genome. Not just in mice, but in humans and other mammals too. It was at this point that imprinting moved from being a mere genetic curiosity to a phenomenon with medical importance. Like the little and large chromosome 7 mice, problems with imprinted genes in a region of human chromosome 15 cause opposing conditions. One is Prader-Willi syndrome, where a loss of the genes that are normally active when they come from Dad leaves sufferers with an insatiable appetite and underactive metabolism along with numerous other health problems. The reverse situation, Angelman syndrome, is caused by missing or faulty versions of genes coming from

Mum, with affected children unable to feed and failing to thrive.

These findings have spawned more than two decades of detailed research aimed at figuring out how the imprinting process works in exquisite molecular detail. Azim, Anne and many others around the world have identified stretches of DNA that are important for genes to pick up imprints as they're packaged into egg or sperm cells. They've tracked down non-coding RNA molecules and proteins that play a part. They've mapped histone marks and DNA methylation in ever finer resolution, and hunted down the modifying molecules that add or remove them. Yet despite all this effort, when I look back on the research field I left a decade ago I see a collection of exceptions. Different imprinted genes seem to be controlled in subtly different ways, even within the same species. Some genes are only imprinted at a very particular point in life or in a specific tissue, but are active from both Mum and Dad's versions the rest of the time. This makes it difficult to draw up hard and fast rules for how it all works. And for families afflicted by imprinted gene disorders, there are still no cures.

It's not just the how of imprinting that is still an open question – there's also the why. Although I've often been advised to avoid asking 'Why?' questions in biology (because the answer is usually 'It just evolved that way') the scientific world is still baffled by the existence of imprinting. For a start, as Azim explains to me, it doesn't make sound genetic sense. 'I have to admit that this is something that I've not resolved in my head. Why you would give up a perfectly safe system where you have two versions of a gene, so if one goes down you have a backup?'

In an attempt to provide an answer, people have come up with all kinds of ideas. One plausible suggestion is that imprinting controls a battle between the sexes that takes place in the developing embryo. This theory is based on the observation that many genes that are only active when inherited from Dad tend to encourage babies to feed lustily and grow big, while Mum's genes have the opposite effect.

The idea is that if a male gets lucky with a lady, then it's in his interests to equip his offspring with genes that make them grow big and strong, as he doesn't know if he'll get another chance to do it. But the female in the deal wants to keep her genetic options open, using her imprinted genes to rein in this overgrowing tendency in case she has babies by a different partner in future (or, in the case of some species, multiple offspring in the same litter by different fathers).

One famous just-so story in the field tells of two closely related species of deer mice – one faithfully monogamous, the other wildly promiscuous. Based on the 'battle of the sexes' theory, the promiscuous mice should have imprinted genes, while the monogamous ones should have no need of them. Reality, however, is not so neatly obliging. The happily married mice have imprinting as well as the serial shaggers. I ask Azim if he has any clues. 'I was giving a talk the other day and people are still asking me why we have imprinting. The truthful answer is that I don't know. But I *do* know that Anne has some ideas.'

❧

Now head of the genetics department at Cambridge University, Anne Ferguson-Smith is one of my favourite scientists. Softly spoken yet fiercely opinionated, with a penchant for smoking foul-smelling roll-up cigarettes (at least, when I first knew her), she's a good person to ask about the why of imprinting.

When I pop in to visit her, everything is in chaos. Originally constructed for the Department of Agriculture in 1910, the building housing Anne's academic empire is crumbling. During the extensive renovation work to restore it, she's been relegated to a temporary office in the far reaches of the even older physiology department. We wind our way up to it via wide stone stairs, clutching the dark wooden banister stained by countless grubby student hands and inhaling a faint whiff of formaldehyde leaking from the ageing teaching labs. As we go, I chatter about some of the things I've found out on my travels – cats with thumbs, fish with hips, wobbly worms and more. In return, she tells me about her recent holiday.

'I just got back from a trip to Thailand, where I stayed in a place that has one of the best spas in the world. I had a series of massages by the veteran masseuse in the spa, who had six fingers on her right hand – she had a second thumb. I've had a lot of massages all over the world, but this woman was the best I've ever had and I think it had something to do with the fact that she had this extra digit. Now, she didn't have full motor control but it provided this extra appendage that influenced the quality of what she did. So sometimes these genetic anomalies shouldn't be considered as negative things, because under certain circumstances they can be positive things.'

Speaking of genetic anomalies, we finally reach a bleak, almost prison-like room at the end of a dusty corridor, and we sit down across a large wooden table to talk about imprinting. In Anne's mind, the 'dosage' of some genes is critically important. For most genes, it's fine to have them firing off messenger RNA from both copies (Mum's and Dad's). But a subset of genes only needs half as much message to be made. In the case of imprinted genes, rather than adapting some nearby control switches to dial down the activity from both Mum's and Dad's versions, evolution has solved the problem by simply turning one off entirely. 'I think imprinting is a mechanism to control the dosage of certain genes,' she says, placing her two index fingers on the table, as if they were a pair of matching chromosomes. 'It's all about dosage – a way of getting different levels of different genes in different places at different times.'

She tells me about a small cluster of imprinted genes she's working on, which are all involved in keeping baby mice warm after they're born. Together, they encode a bunch of proteins controlling the animals' metabolism and fat-burning, generating vital heat to keep them toasty once they're big enough to start straying from their Mum and feeding themselves. 'We can breed mice with a mutation that affects the dosage of the genes in this cluster. If you have a tiny bit too much of one of them, it slows down their fat development for a few days. And if you have a tiny bit too little of another one, it interferes with the hormones that control their metabolism. Together, it means that these pups can't regulate their body temperature when it's time for weaning and they die. Every single one of them dies.' But, fortunately, it doesn't have to be this way. 'If we put them on a hot mat for the length of time it takes their fat development to catch up, then we can save them all. That's two completely different genes in two different biological pathways, but they all come to the same point – controlling the organism's adaptation to independent life. And because they are both exquisitely dosage-sensitive, they are located next to each other and regulated together, so the levels can be controlled. And I think that's why genes are imprinted.'

To take her ideas further, Anne's been hanging out with the National Evolutionary Synthesis Center, a big evolutionary biology centre in North Carolina in the US. 'So far there's nothing – *nothing* – that really explains imprinting. We had a big meeting in the fall and it was clear that we don't know anything about its evolution.' 'But what about the promiscuous mice?' I ask. 'What about the battle of the sexes?' 'It's not relevant,' she says, sternly. 'When you were working with Azim it was all about the growth of the embryo and the placenta. But now it's an awful lot more about postnatal adaptations, about metabolism, about the brain, about behaviour. It's about the transition to independent living and leaving the nest. It's about these interactions between a mother and her offspring, which are hugely important for survival. That fat gene I was talking about earlier? It's just a 30 per cent increase in activity that leads to the pups dying. That's nothing. It's an exquisitely dosage-sensitive function that controls the whole life of the organism, and imprinting regulates that.'

I'm not sure I'm entirely convinced. To me it doesn't seem like a very robust system to have the life of an animal depend on such a tiny change in gene activity. But Anne is having none of it. 'It's because the gene does different things at other times. It's also involved in making new brain cells in the adult brain. The dosage there is really important too, but you need twice as much. If you knock out one copy then you can't make brain cells.' Suddenly it makes sense. Although it's a risky strategy, harnessing the power of imprinting to make changes in the level of gene activity at particular times and in particular places is yet another way that evolution has enabled our genome to get more bang for its buck. Like the other adaptations we've encountered so far along this journey, such as RNA splicing and editing or flipping different control switches, imprinting allows us to use our limited repertoire of 20,000-or-so genes in ever more inventive ways.

For now, imprinting remains a biological and medical curiosity. Some researchers like Anne are still beavering away, trying to figure out why it happens and exactly how it works, but others just chalk it up as 'one of those things'. But although it's certainly strange, imprinting isn't the *really* weird stuff.

CHAPTER TWENTY-ONE

Meet the Mickey Mouse Mice

Fortune favours the prepared mind.

Attributed to Louis Pasteur

Parc Valrose, the home of Nice University's science campus, sits proudly on a small hill above the elegant hotels and chic chalets clustered on the Cote d'Azur. In its previous incarnation it was the winter residence of Paul Georgevitch von Derwies – self-styled King of Russian railroads – who played host to Queen Victoria when she popped in from her palace round the corner. In 1965 it was turned over to more academic pursuits. The sumptuous château at the top of the hill has been converted into offices for the boss and his admin staff, while the researchers labour in crumbling concrete blocks against a background of interminable building work.

Although royalty and Russian railway barons may have once wintered here, Nice has become a distinctly summery destination. Woefully out of season for my visit, the December drizzle stings my face as I trudge up to the campus along streets paved with ennui and dogshit. I'm here to see Minoo Rassoulzadegan, or – as she's sometimes referred to by other people I've talked to – 'that woman in Nice with the weird mice'.

That woman herself is neat and friendly, with an effortless put-togetherness and a slash of bright lipstick. In a tidy historical arc, it turns out that Minoo's long-time mentor and office-mate François Cuzin was François Jacob's first graduate student at the Pasteur Institute, just after he and Monod figured out the components of the first genetic switch. Now advanced in years, Cuzin still comes in daily to plan experiments and produce papers – or, as I later discover to my pleasant surprise, to take visiting science writers out for long, boozy lunches and a tour of the campus. 'So I guess that makes Jacob your scientific grandfather?' I ask, when she reveals her pedigree. 'It is not bad, eh?' she says proudly, in heavily accented English. 'Sometimes François gives a very big compliment to me when he says "Ah, now you are thinking like Jacob!"'

Now in her sixties, she's been in Nice since the age of 18 after leaving her native Iran. As we walk up the shabby stairs to her office, we talk about our shared love of early mouse embryos. Those delicate baubles floating in the liquid universe of a Petri dish are perhaps the only thing I really miss about lab work. I'm surprised to discover that she still does nearly all her experiments herself, rather than farming them out to post-docs, technicians or students. 'I still do everything by myself, by my own hand,' she explains, settling me into a chair in her small shared office and fussing with one of those posh little coffee makers with the foil capsules. 'Most of my ideas come when I do experiments myself because you see better the biology and you know why it does or doesn't work. Sometimes there is an idea floating – often these ideas are around but the experiment tells you something even more interesting.'

Minoo certainly noticed something quite odd while she was busying herself in the lab. But luckily, she says, para-phrasing Pasteur, 'My mind was prepared to see it.' It all started from a completely unrelated experiment. In the late 1990s she was busy creating genetically engineered mice that could turn certain genes on or off in the developing egg and sperm cells. One of Minoo's targets was a gene called Kit.* It encodes a protein that sits on the surface of cells, receiving signals from the world around it that tell particular genes to be switched on or off. Like many important signalling genes, Kit has a lot of different jobs depending on when and where it's active. These range from telling cells in the early embryo to become germ cells (the precursors of eggs and sperm) to controlling the pigment cells that make skin and hair dark or light. And, as is

---

*We've met Kit already, or at least the protein that binds to it – Kitlg – in Chapter 7, along with the blondes, fish hips and penis spikes.

also often the case with many signalling molecules, faults in Kit are found in various types of cancer.

Minoo's experiments were designed to generate an entire litter of mouse pups all carrying one faulty copy of Kit and one healthy one. Animals can't develop in the complete absence of Kit, so Minoo had to make do with knocking out just one copy and leaving the other as a backup. Regardless of any other genetic flaws they might have had, these animals were certainly distinctive, if not downright attractive. Well, for mice anyway. 'They all have these little white feets! They are very cute – I can show you a picture ...' She flicks through a presentation on her computer and alights on a charming photo of a chestnut-brown mouse bedecked with four white paws and a bleached tip to its tail. Straight away I spot a resemblance to another famous rodent. 'They look like Mickey Mouse with his gloves!' I gasp. 'Yes, I suppose they do,' she laughs.

She was about to dismiss the gloves and socks as a mere curiosity when things started to get strange. According to Mendelian dogma, when Minoo bred two of these Mickey Mouse mice together a quarter of the animals would inherit two faulty copies of Kit, so they couldn't survive. Of the surviving pups, roughly two-thirds of them should have a single faulty Kit gene plus a normal copy, along with the expected white gloves, socks and tail. The rest should look completely normal and brown, as both their Kit genes are healthy. Weirdly, every single mouse had white feet. Yet when she tested their DNA, only two-thirds of them carried a faulty Kit gene. The rest were completely genetically normal. Yet there they were, flaunting their Mickey Mouse gloves around the cage. 'I thought, this cannot be! It is not possible! Mendel says it is not possible!' She throws her hands up in the air in a fabulous gesture of despair and bewilderment.

I know I have spent most of the past few chapters explaining how organisms can carry a faulty version of a gene but look perfectly fine, but one thing in genetics should always hold true: if you have a normal version of a gene, it should do what it's meant to do and give you a normal phenotype. But these

Mickey Mouse mice, who carry two normal versions of the Kit gene, look for all the world as if they are carrying the faulty version. That's not just a bit weird. That's a totally messed-up version of biology.

'Did you just look at this and go … huh?' I ask. 'I thought I had made some mistake. So I did it all again, and I observed the same thing. Then I started to think, it's not possible that I'm doing this wrong every time. So what is going on?' To try and figure it out, she kept breeding the mice together. And it got stranger. If she bred the Mendel-defying Mickey Mouse mice (with two healthy Kit genes) with perfectly normal mice, the gloves vanished after a few generations. But if she kept crossing brother and sister Mickeys together, things got more extreme. The more the mice bred, the more white they became. White blotches appeared on their backs or other bits, and some were born completely white, even if they had two normal copies of Kit and should have been boringly brown.

Like any scientist with a strange but exciting discovery, Minoo couldn't wait to speak about her unusual begloved mice at conferences. But it was just too weird for most people to cope with. 'I started to talk about it and everyone said, "Oh, you will never be able to explain this. You will never, never explain it!" But I felt I had touched something.' It was then she had a brainwave. Although the peculiar Mickey Mouse young themselves had inherited two normal copies of Kit – one from Mum, one from Dad – the egg and sperm that made them were created in the body of a parent that *did* have a faulty Kit. So was something else other than just DNA being transmitted from parent to pup, which was having an effect on the babies' normal Kit genes?

Given that the effect worked both ways, regardless of whether the original Mickey Mouse parent was male or female – and bearing in mind that eggs are packed with all kinds of protein and RNA goodies, making it hard to start unpicking what's going on – Minoo began by looking at sperm. Because sperm cells are so tiny, barely more than a tightly compacted ball of DNA with a tail, there's long been a somewhat sexist idea in developmental biology that males contribute nothing but their DNA to the next generation. 'People said oh, even if there is something else coming from the sperm then there is very little of it and it doesn't do anything!' She flaps her hands dismissively at the idea. 'So I went to have a look.'

Using a technique called electron microscopy, which can zoom right down into cells and reveal individual molecules, Minoo peered closely at the squiggly tubes inside a male mouse's testicles that hold their sperm. What she saw there was RNA. In fact, contrary to popular opinion, there wasn't just a little RNA lurking inside – there was *loads*. But just discovering truckloads of RNA in a mouse's balls doesn't prove that it's responsible for passing on the Mickey Mouse trait. More evidence was needed. To find it, she prepared RNA from sperm from animals containing one faulty copy of Kit. Next, using a tiny glass needle and steady hands, she carefully injected it into fertilised egg cells that were completely genetically normal and had never been near a Mickey Mouse. Then she replaced the manipulated eggs back in a mother mouse's womb. To her delight, they grew into pups with strange white socks, gloves and tails.

It wasn't just the sperm RNA that had this near-magical effect. RNA from the brain, where the Kit gene is also active, worked too. But, somewhat worryingly, RNA from genetically normal mice very occasionally resulted in Mickey Mouse gloves in the manipulated offspring, which were themselves also genetically normal. Confusing, huh? Perhaps not.

'This result led me to think there is something else going on. I wondered if maybe what's happening in these injections from normal mice is that there are some small fragments of degraded RNA made from the Kit gene. And that led us to micro-RNA.' As you'll recall from Chapter 14, micro-RNAs are short fragments of RNA apparently blessed with a whole gamut of gene-regulating duties. What if the faulty Kit gene was making a micro-RNA that could get into sperm, pass on to the next generation and somehow turn down or switch off

Kit in the resulting embryos? If that happened right at the start of development, when the future mouse was just one fertilised cell, then it would have the same effect as knocking out the gene completely. And if the normal RNA from a complete Kit gene got a bit shredded as Minoo injected it, then that might accidentally generate a micro-RNA that could have the same effect too.

Taking advantage of the ability of companies to quickly and cleanly make any small RNA to order, she set about buying in a couple of micro-RNAs that were already thought to have an effect on Kit, based on similarities between the sequence of the RNAs and the Kit gene, and injecting them into mouse embryos. Again she saw the same thing. Adding the Kit micro-RNAs to genetically normal fertilised eggs was enough to create Mickey Mouse pups, proving in her mind that it's these little fragments of RNA – transported in sperm from one generation to the next – that are leading to a dramatic and heritable effect on the Kit gene.

Many other geneticists are sceptical of her work, and I have to admit it is a pretty crazy story. But it would be easier to dismiss out of hand as a piece of intriguing but inexplicable phenomenology if she hadn't then gone and done it again. Minoo's second strange discovery was another piece of Pasteurian serendipity. Like any diligent scientist, she also tried a control while she was testing the Kit micro-RNAs. This was a random fragment known as micro-RNA 24, which had no similarity in its order of RNA letters to the sequence of the Kit gene. If this control fragment couldn't create Mickey Mouse mice, then she would know for sure that the effect had to be caused by the micro-RNAs with Kit-specific sequences rather than any old micro-RNA. As a negative control, micro-RNA 24 worked perfectly. No white gloves or socks to be seen. Not even a whisper of a blonde tail tip. Yet there was something else unusual about the recipients.

'They are like super-mice! Oh my God, they are bigger and they stand up much earlier on their feet. Straight away you can see it. If you put a normal mouse on a little platform at two weeks of age they just walk along and fall off, because they don't understand it yet.' She paddles her hands floppily in front of her like a baby animal tripping over a step. 'But these big ones,' she says, curling her fingers under her chin as if peering over a wall, 'they look … they come back. So they are smarter as well.'

The obvious next question is: what's the identity of this amazing super-mouse gene? 'At the time the target of the micro-RNA was not known. We had a look for it, and we found it was a gene called Sox9.* But unlike other cases where you see gene silencing, we discovered that Sox9 was *more* active in the animals we had injected with it.' This is kind of strange, especially since the Mickey Mouse gloves are the result of a gene being shut down by the micro-RNAs. But it's certainly in keeping with the wide range of (and in some cases contradictory) biological effects that these tiny molecules seem to be capable of.

So that's two examples. Any more? 'Around this time it was the beginning of the discovery of micro-RNAs, and there were a lot of presentations at meetings about which micro-RNAs are found in what organs. So I saw that one particular micro-RNA is only in the heart muscle, and they say it is very important in the development of the heart, in its physiology.' Recognising that her specialist knowledge of the heart was somewhat limited, Minoo tracked down a cardiologist colleague who would be able to spot if this third micro-RNA was having any effect. In the end, it turned out that the result was so striking that even she couldn't have missed it. Either she had struck it lucky for a third time, or a pattern was starting to emerge. 'You can just see it! The animals that have the extra micro-RNA, their hearts are beating much faster. And the interesting thing is that the number of cells in the heart is the same as in normal mice, but the heart is larger.'

---

*Confusingly, the Sox9 gene does not give mice white socks.

If a doctor saw this kind of problem in humans, they'd call it 'cardiac hypertrophy', which translates as heart overgrowth. Extensive DNA studies have so far drawn a blank on the specific genes involved, but they've often shown that a gene called Cdk9 is overactive in people with the condition. And guess what the target of Minoo's heart micro-RNA turned out to be? Cdk9.

Given what a huge problem heart disease is in humans, I want to know what happens to these mice. 'They're OK,' she reassures me, 'but their hearts beat faster and the muscle cells are all disorganised, like in human patients. Here we keep them in good condition in the laboratory, but in nature who knows? Maybe suddenly they have to run and then they have a problem.' She mimes the scampering action of a mouse running along then keeling over dead. I giggle at her pantomime, but it's a deeply serious problem. Every single week in the UK 12 apparently fit and healthy young people drop dead from an undiagnosed heart condition. Maybe, she thinks, some of these tragic deaths are due to rogue fragments of RNA that came along for the ride with a sperm, causing Cdk9 to be overactive and messing up the heart muscle.

It's interesting that genetic studies have consistently failed to produce strong candidates to explain a significant proportion of sudden heart failure, or any other kind of heart disease. There are a few major suspects – like the Titin gene we met earlier – but little else of note, and certainly not enough genes to explain every case. It's a similar story with large-scale studies for a wide range of complex conditions that are thought to involve subtle variations in many different genes, including diabetes, cancer, autism, depression and more. Faults in genes such as BRCA1 and BRCA2, which are relatively rare in the population but have a big impact on breast, ovarian and prostate cancer risk, are a small part of the picture. Most of the gene variations linked to any particular condition contribute, at best, a few percentage points to the overall risk. And yet we know that many of these diseases – as well as other complex traits such as intelligence – must have a significant inherited, and therefore presumably genetic, component.

The hope is that all these tiny fractions of genetic variations linked to small increases in risk will eventually add up to the full picture. But as researchers undertake ever more detailed gene-sequencing studies with more and more people, this 'missing' genetic contribution (known as missing heritability) is still stubbornly refusing to appear. However, what if there's a whole part of the puzzle that's not inherited in the DNA at all?

'Everybody knows that diabetes is a big problem for society,' Minoo says, moving to another part of her presentation on the computer. 'Studies show that if people eat a high-fat diet they can become obese and develop diabetes. Also this can transmit to the next generation. This is not just in people but also in rats, in mice. But what is not known is how.' Most people in the field have pinned their money on classic epigenetic tags such as DNA methylation or histone modification being responsible for the phenomenon, but there's scant evidence so far that these markers are the true culprits. As you're probably expecting by now, Minoo is going all-in for RNA. 'In this case people have been unable to show experimentally that it is histone modification or whatever – the usual suspects. So it became obvious to me to take RNA from fat mice and see what happens.'

She taps a neat fingernail on the monitor, talking me through the experimental protocol. 'So we take some mice and feed them a diet that is 20 per cent butter …' 'Wow! Do you just constantly feed them croissants or something?' 'Haha, *non*! We just add the butter to their food, and in three months they become obese and diabetic. So the question was, this fellow here …' She points at a chubby butterball of a mouse in the photo on the screen. 'What is in his sperm? So we take the RNA from the fat males and we inject it into one-cell embryos. Then we ask, are these pups also fat?'

The answer was yes. And what's more, when these fat babies grew up they also produced fatter pups. None of this

happened when Minoo tried RNA from male animals fed a
normal diet. But here's where things get a little complicated.
Unlike the neat examples of Kit, Sox9 and Cdk9, there
doesn't seem to be one specific micro-RNA involved in this
hereditary pattern of plumpness. Or if there is, she hasn't
found it yet. 'It's a very complex condition, and you probably
need a group of genes. We have found there are a few micro-
RNAs which increase in activity under a high-fat diet. I have
tested two of them, and with one there is no effect, but with
the other you do see something. They are definitely bigger,
and if you breed them with a normal partner, in the next
generation they are still a bit fat.'

These experiments provide a tantalising hint that RNA
hitch-hiking in eggs or sperm, rather than DNA methylation
or histone tags, might be the medium for transgenerational
epigenetic inheritance. And maybe, just maybe, it might fill
in some of the missing parts of the inheritance puzzle for
complex traits and diseases.

As I look around the cramped office, I notice a piece of
paper pinned to the far wall. Spoofing a classic film-noir
poster, it's a picture of a painted dame posing in fear at some
unknown terror. The tagline '*The case of the missing heritability!*'
marches across it in mock-handwritten script. Based on her
RNA findings, Minoo now has big plans to track down this
elusive villain. 'I am setting up a new lab in Turkey,' she tells
me. 'My family are in Azerbaijan so it is closer for them to
visit me. But although my genotype is Iranian, my phenotype
is Niçoise – my heart will always be here. Yet Turkey is
investing heavily in science, and here in France the situation
is not so good any more.'

She plans to use her new base to study autism. It's an ideal
candidate, as it's a complex disease with undeniable links to
a huge number of different gene variations – hundreds of
'genes for autism' have now been found – but very little
understanding of how it is caused or inherited. 'I have found
families with a lot of cases of autism due to cousin marriage.
They are very inbred. In one family there are three
generations affected, and by the third generation 6 in 10 of

the kids have it.' By keeping it in the family, these people are inadvertently mimicking Minoo's strange Mickey Mouse mice that become increasingly piebald when they breed with each other. She's currently working on developing a mouse version of the autism family, trying to track down micro-RNAs that might be involved. And, if she can manage it, searching for them in humans, too. 'One father in the family, he has autism and schizophrenia. He is very difficult to work with. I wanted to get a sample of his sperm, but he would not co-operate. So I asked his wife. Later she came back to me with a small pot. I did not ask how she managed it.'

It all comes back to the idea of the embryo as a singularity: a unique point in time when an organism exists as one solitary cell with one complete set of DNA. Each cell in the germline is the same: a single cell (this time with half a set of DNA) that might one day be the lucky egg or sperm to make a baby. Any epigenetic alterations to these cells that can be copied and passed on as the cells divide and grow will create ripples that can last a lifetime, or even further. It seems almost magical. 'It is not magic,' she says. 'There is a molecule!' According to Minoo, tiny fragments of RNA can have a potent effect on the unique characteristics of an individual, starting from the egg and sperm that make them. She believes that small, specific pieces of RNA in the egg or sperm might semi-permanently alter the activity of their target genes in the resulting embryo, starting from the very earliest point in development and in the absence of any underlying inherited genetic alteration. And because the precursors of eggs or sperm, the germ cells, are laid down shortly after fertilisation, they might also experience the knock-on effects of the RNA, so the phenomenon can affect the next generation too.

There's still a lot of speculation here, but this peculiar non-genomic inheritance – passing on characteristics without passing on DNA changes – flies in the face of everything we think we know about genetics. Darwin was wrong, Lamarck was right! Suck it, Charlie! In fact, Darwin might have been right about this too, although for 150 years everyone thought he *was* wrong.

❧

Back in the 19th century, before anyone had come up with the concept of genes or identified the molecule responsible for encoding them, there were plenty of ideas about how heredity might work. Charles Darwin's personal favourite hypothesis was the concept of 'gemmules', which he outlined in his book *The Variation of Animals and Plants Under Domestication*. These were proposed to be small particles of inheritance that travelled around in the bloodstream, gathering information about the characteristics of the creature whose veins they traversed. Then somehow when animals mate (the details here are sketchy …) these gemmules combine in the embryo to create all the structures necessary for life, from legs to liver. And because one set of gemmules comes from each parent, this neatly explains how the particular traits of Mum and Dad can be blended together in their offspring.

Francis Galton, Darwin's brilliant but uncomfortably racist cousin,* loved this idea and set about trying to prove it right. Surely, he thought, if gemmules were circulating in the bloodstream, then a blood transfusion from one animal to another should result in at least some of the characteristics of the donor being passed on to the recipient's offspring. His organism of choice for these experiments was the bunny rabbit, and enormous numbers of animals were bled and bred over three years in an ultimately fruitless search for these mysterious packages.

---

*Galton is often remembered as the founding father of eugenics, and was obsessed with the idea of trying to breed better, smarter humans. But he was also an imaginative scientist who could count among his many inventions the 'gumption reviver' – a sort of mobile dripping tap invented while he was an undergraduate, designed to sit over his weary head and keep him alert.

Of course, once people started to figure out the true nature of heredity, all talk of gemmules went out of the window. But I've been hearing them mentioned – albeit in somewhat hushed tones – by several of the researchers I've spoken to on my journey to the frontier of genetics. When I say the word to Minoo, she becomes ecstatic. 'Ah, I'm a big fan of Darwin! I used to read him everywhere, at home, on the plane, on the beach, everywhere! I always thought, why should nature make it so that acquired characteristics cannot be transmitted? This is stupid if you lose all the experience acquired in your life. And then I read in Darwin about the gemmules.'

Could small fragments of RNA packaged in eggs and sperm be modern-day gemmules? There's a long-held view that eggs and sperm, and the germ cells that produce them, are completely and impenetrably separate from the rest of the body. But the heretical discovery that characteristics acquired in life *can* be passed on in some form proves that this can't be true. Recently researchers have discovered RNA fragments moving around the body in little packets, called exosomes, which can be taken up by cells and affect gene activity. This kind of mechanism would provide a way that the rest of our body can 'talk' to eggs and sperm, passing messages on to the next generation. There's a long way to go to prove this actually happens in animals, let alone humans, but there are a few intriguing hints that it might. For example, an Italian researcher has found that RNA from human cells transplanted into mice can end up in their sperm, although it is a highly artificial experimental system and might not reflect reality.

Maybe these exosomes are acting like Darwin's gemmules, gathering information about how cells are responding to changes in the environment from around the body in the form of RNA and depositing it in stored eggs and sperm? Maybe they're not. But it's certainly a fascinating idea to ponder. It's still very left-field and controversial, but if the principle holds true we're going to need a major rethink of our conventional understanding of genetics and heredity.

Someone who thinks this change is coming is Oliver Rando at the University of Massachusetts campus in

Worcester. He's just received a grant for the best part of a million dollars from the US National Institutes of Health to figure out if diet can affect levels of particular RNA fragments in sperm. His hunch is that complex diseases like diabetes, which are influenced by a seemingly impenetrable mix of environmental and genetic causes, might be down to transgenerational transmission of RNA. And Minoo herself thinks that up to a third of our genes might be affected in some way by inherited fragments of RNA. At the moment this idea, like Minoo and her Mickey Mouse mice, is lurking at the scientific fringes. But that might be changing.

'We're trained to think about genetics in a certain way,' she explains. 'At first you don't accept something like this, but you have to open your eyes to it and accept the variation. Some scientists, the good ones, are ready to accept, and they ask me for more data. But some people say, "Oh, nobody can reproduce this."' But now other labs *have* reproduced her findings, or at least found strong hints that RNA might be having an influence across the generation gap. 'Some people forget to mention my name or my papers in their research. But I think it's OK because the science is real, and there are other people doing it now.'

I have a feeling that this particular area of research is going to become as colourful as the glorious sunset over the Bay of Angels that evening as I head back to the airport. Watch this space.

CHAPTER TWENTY-TWO

In Search of the 21st-Century Gene

One Monday morning in midsummer I take the train to Leeds, heading for a small academic conference. My aim is to speak with Evelyn Fox Keller, grande dame of science history and philosophy, groundbreaking feminist and author of many highly respected academic works. Now pushing 80, she's the keynote speaker at the meeting, which brings together sociologists, historians and philosophers to discuss the nature of genetics.

I arrive with some trepidation. These are not my people, and I'm feeling a bit out of my depth after the first day's talks. I'm much more comfortable with the practical banter of lab scientists than with highfalutin philosophising. Adding to my problems, a bunch of academic historians then take me to the pub to experience the twin joys of cheap local ale and intellectual discourse into the small hours.* They also gift me with a suitably repentant hangover the next morning.

When I finally make it down to the breakfast hall I notice Evelyn sitting alone, small and thoughtful. 'I'm writing a book about how genes work,' I tell her, settling myself in the chair next to her and hoping I don't still smell of beer. 'Do you have any advice for me?' 'Well, for a start I think you should try not to use the word gene at all,' she tells me sharply, fiddling with her coffee cup. 'I'm not sure my publisher would like that,' I reply. She raises an eyebrow at me, like a teacher listening to a poor excuse for lost homework. We sit in silence for a few minutes as I try to force down some limp toast.

---

*I'm sure at some point someone was talking about 'The inherent sexist violence in *Thomas the Tank Engine*', but I may have been imagining it.

Finally she clears her throat to speak. I am instantly, urgently attentive. 'Do you know if I can get any more coffee?'

I dutifully trot out to the drinks machine, feeling disappointed that our encounter isn't going better and wondering if it would be rude to ask her for aspirin. But when I get back with my caffeinated offering, she tells me a story. 'I went to a conference where a guy was talking about all these new genetic variations they were finding for different diseases. And so I stood up at the end of his talk and asked him how many of them were actually *in* genes. He said "Er, very few". And so I said, "Well, how can you call them genes?" And they all went crazy.' She takes a sip from the cup and looks at me again, eyes glinting mischievously. 'What did they say?' I ask, imagining the furore as this little old lady told them off. 'Once the hubbub died down, one scientist stood up and said in a quiet voice, "But I *like* genes."'

🐾

This is the problem that we face when we try to think and talk about how our genes work and what they do. We all *like* genes. We like headlines that tell us when researchers have found the latest 'fat gene' or 'cancer gene'. Scientifically speaking, we like the simple idea that a gene makes a protein that does a thing in the body. And when the gene is faulty it means the protein is broken and doesn't work, which causes a problem. Yet, as I hope I've managed to explain, it's a lot more complicated than that.

Statistically speaking, almost none of our genome encodes for proteins. But a significant and hotly disputed fraction of the rest of it certainly does something important in terms of controlling when genes are turned on or off, being read into RNA, providing fodder for evolutionary processes, organising the three-dimensional structure within the nucleus, and almost anything else you can imagine. We have banks of control switches for our genes, zombie genes, long and short non-coding RNAs, miniature Smorf genes, genes within other genes, alternatively spliced and edited genes, jumping genes, imprinted genes, wobbly genes … The list goes on and will doubtless continue to grow in the future. All of these things add up together to make us the person we are. But are any or all of these things actually, y'know, *genes*?

This might not seem like such a big deal, but clear definitions are very important when trying to communicate concepts in genetics, whether to researchers, to students or to the public. I've certainly got more and more confused about how to think about genes the further I've got through writing this book, and I wouldn't be surprised if you're also feeling a bit bewildered after reading it. That's because it's really confusing. And, to be honest, it turns out that even scientists don't really know what a gene is.

Someone who's been exploring this problem is Karola Stotz, a philosopher of science at the University of Sydney. I stay up late one freezing night in January to speak to her at home in Australia. As the slightly pixelated Skype picture flicks into view, I feel a pang of jealousy at the scene as antipodean summer sunshine floods into the screen. She talks animatedly in a strong German accent, her hands constantly waving, while an unseen animal or bird chirrups occasionally in the background. I begin our interview by coming right out with it. 'I'm not even sure what a gene is any more, so I thought I would ask you.'

'We have worked a lot on this question, starting 14 years ago,' she tells me, her dangling silver earrings flapping distractingly against her neck. 'We were doing a large questionnaire with different kinds of biologists – molecular biologists, developmental biologists, evolutionary biologists … basically, we wanted to know how they understand genes in their work.' Her survey involved giving the scientists a range of different genetic scenarios from nature. Not just the easy stuff like straight-up protein-coding genes, but the crazy things. Overlapping genes, genes built from stretches of DNA on two different chromosomes, transcripts running in opposite directions through the same sequence and more. 'There are so many weird things happening in the genome, so I found 12 of these odd cases and gave them to the biologists to think about. I asked them for each case – is this one gene, or more than one gene? And if more than one gene, how many genes are involved?'

In fact, in many of these situations there's no right answer. But you wouldn't know it, from listening to the replies. 'The weird thing is that no one admitted to not knowing. Nobody was saying, "All bets are off, one gene or more than one gene, what do I care?"' she tells me, shrugging her shoulders as her earrings wriggle and dance. 'They either committed to saying it's one gene, or it's definitely more than one gene.'

Interestingly, the answers they got from the researchers were strongly tribal. Evolutionary biologists tended to favour certain answers, while molecular biologists preferred other definitions. Effectively, scientists don't really adhere to a strict definition of what a gene is – they just use the one that's convenient to them for their work. Although this may sound like philosophical noodling, once Karola started doing more in-depth interviews with researchers, she realised that many of them were intrigued by the questions she was raising. 'Suddenly they had to think about it, and they found it quite interesting. They said things like, "Right now I'm working on *this*, and therefore for me a gene is *that* – but one day I may choose a different field and then the gene may be something else for me." So I think that scientists can be very pluralistic about what kind of gene concept they're using. There are at least three scientific representations of the gene around at the moment, depending on what you're using it for.'

If there's so much flexibility in how scientists themselves refer to genes, what does that mean for the rest of us? I ask Karola how she herself sees our genome and the genes within it. 'For me it's just a Lego box of developmental resources.' She pronounces it 'Leego', delving her hands into an imaginary crate of the stuff. 'The bricks are the possible templates to make a protein. But only about 1 per cent of our genome are these Legos. The rest are regulatory regions, which are the instructions for what to do with them – how to fit them together differently, how to make the most out of it.'

She thumps the table for emphasis and the picture jiggles. 'Literally, you are a person who has this wonderful Lego box, and you can do a whole lot with it.'

In their book *Genetics and Philosophy*, Karola and her co-author Paul Griffiths highlight what I think is a rather charming phrase: genes are the things you can do with your genome. Personally, I'm coming back to an even simpler definition of a gene as just being an inherited thing that does a thing. Inheritance is a key point here. DNA in a test tube is not alive. RNA and protein molecules are not alive. Scientists can mix RNA polymerase and a bit of DNA together and see RNA being written from the genetic template. Adding a different polymerase creates a new copy of the DNA instead. But this isn't life. In 2010, gene-sequencing wizard Craig Venter revealed 'Synthia' – a bacterium created from a string of DNA code that had been glued together in the lab and put into a cell from which the original DNA had been removed. Predictably, the media lost its collective shit over the announcement, with headlines excitedly claiming that this was the first 'man-made life form'.

It wasn't. Although an impressive piece of genetic engineering, Synthia still needed the biochemistry and structure of an existing cell to make her genes work. However clever Venter might be, he can't add DNA to a test tube of molecules and make it live (at least, for now). Every cell, from the simplest bacterium to the billions making up our own bodies, is alive because it came from another living cell. Doesn't matter whether it's the boring budding of one yeast cell into two, the hayfever-inducing pollination of plants, or the sweaty entanglement of two human bodies bringing egg and sperm together, life is reproduction and reproduction is life. This means that it's too simplistic to think about what our genes do and how they work without including the context of inheritance and reproduction. Making a baby isn't as simple as shoving two naked sets of DNA together. As well as bringing their half of a genome that has been shuffled, invaded, tweaked and mutated over millions and millions of years, egg and sperm each come with their own biochemical baggage. Not only do transcription factors and epigenetic marks get pre-loaded onto DNA in germ cells but – as we saw from the Mickey Mouse mice – RNA and potentially other molecules might play a part in influencing gene activity across the generations.

Then there's the environment to add into the mix. No living thing exists in a void entirely on its own, and whatever that environment looks like – the molecular gloop surrounding your cells, the womb you grew in, the family that nurtured you, the home you live in, the food you eat, the air you breathe and everything else – billions of years of evolution have provided cells and organisms with an impressive array of mechanisms for responding to changes in the world around them. How much of this is strictly encoded in the DNA is still a matter of heated debate, but ultimately it's the DNA sequence that acts as the template for making the whole zoo of RNAs that we now know about, directs the docking of transcription factors, and signals where epigenetic tags should be stuck.

As we've seen, alterations in DNA – whether big or small – change the way that genes work, with subtle or dramatic effects on the resulting organism. Rinse and repeat over millions of years and we end up with the dazzling diversity of life on this planet. However (and this is an important point), evolution is not a flawless engine, continually driving living things onwards to make better and better versions of themselves in the generations that come after. This misconception comes about as a result of the way that evolution is often described, especially in school. We're taught that a DNA sequence gets altered, creating some kind of change in the resulting organism. If this is advantageous, then these genetic changes get to stay in the population. If they're not, they get ditched. Over time the useful variation might even become the only version of that part of the genome in the whole species, like the missing control switch in the hipless lake-dwelling sticklebacks from Chapter 7. Evolutionary biologists refer to this as 'fixing' a trait.

This process is natural selection at its finest – survival of the fittest and all that. And while this is certainly one way that genes shift and change over time, it's not the only one. Things aren't as clear cut as a school biology class might make them out to be. That's because evolution doesn't really care whether you're the bestest, fittest organism you can possibly be. All it cares about is that you get laid and pass your genes on. You could have a genetic variation that makes you 10 times smarter than everyone else, but if you spend all of your time in the library instead of trying to meet someone of the opposite sex and make babies with them, your brilliant genes will die with you.

All the time, random changes are happening in our genomes that get passed on to the next generation. Some of them are handy, some of them are positively damaging, while most are just neutral – the genetic equivalent of a Gallic shrug. *Bof.* Whatever. Due to the vagaries of life and love, the proportion of some of these neutral variations will increase in the population, despite not being actively advantageous, while others fade away. This is known as genetic drift, conjuring up a nice image of a random tide of DNA variations ebbing and flowing through time, with as little purpose or direction as the sea splashing on the shore.

This vision of genetics is a stark contrast to the analogies that started to come forward in the 1970s in the wake of the twin revolutions in molecular biology and electronics. Everyone was buzzing with the idea of DNA as a form of computer code or blueprint, with genes forming neat regulatory circuits that could be pulled apart and put back together again. Although this view of genetics-as-electronics was helpful in some ways, time has shown that it was deeply flawed in many others. Real life is a long way from mimicking the tidy precision of a pre-printed circuit board. Our genome was not designed by an engineer, plotting out the most sensible way to make protein X or respond to signal Y. It was bodged and pasted together, not needing to be great, but just good enough to make a human that can pass it on.

Probably the thing that has struck me the most is how consistently this process works, despite the fact that it's driven by essentially stochastic (*i.e.* random) chemical interactions. As I learned from people like Wendy Bickmore with her baubles and Ben Lehner with his wobbly worms, when you get right down to the level of the writhing DNA and blobby proteins inside a single cell's nucleus, physics takes over from biology. The interactions between control switches and transcription factors are flaky, and it's a matter of chance that the right things will come together at the right time. Obviously things have evolved so there's a pretty good chance that it'll happen, otherwise we wouldn't be here, but it's still a statistical event rather than a guaranteed one. Engineers, mathematicians and religious believers hate this idea, maintaining that a better understanding of the complexities of DNA will explain exactly how it all works (or that we'll give up and just accept that God did it). But it won't, in my view. Not completely. We live in a world driven by probabilities, not certainties.* All too often genetics is sold to the public as solid fact: this gene does *this*, that gene does *that*. Yet as I've spoken to scientists in the course of my travels, it's become increasingly clear to me that there's a huge amount we don't know about how our genes work. There may well be things we can never know. But the pace of change is accelerating rapidly, and there's talk of entering a 'golden age' of genomics where high-speed, cut-price gene sequencing will finally lay open more of the mysteries in our DNA.

In his speech awarding François Jacob and Jacques Monod their 1965 prize for figuring out how genes work in bacteria, Nobel committee member Professor Sven Gard said, 'Now that we know the nature of such mechanisms, we have the possibility of learning to master them, with all the consequences which that will surely entail for practical medicine.' More than three decades later, the researchers who published

---

*That still doesn't mean that God did any of it, unless you believe the sole attribute of God is statistical randomness.

the first draft sequence of the human genome wrote, 'We find it humbling to gaze upon the human sequence as it comes into focus. In principle, the string of genetic bits holds long-sought secrets of human development, physiology and medicine. In practice, our ability to transform such information into understanding remains woefully inadequate … Fulfilling the true promise of the Human Genome Project will be the work of thousands of scientists around the world.'

I'm not even sure that's going to be enough. We've come a long way since Jacob and Monod first discovered how to throw the genetic switch that turns a bacterial gene on or off, but it feels like we're barely scratching the surface in terms of understanding living systems any bigger than bugs. And as soon as things get to the kind of level of complexity that's required to build a human, the data and computational analysis required to make sense of the genome and how it works becomes mind-boggling.

And then there's the question of what is 'the human genome' anyway? You may imagine that this should be a done deal by now. Surely, you might think, we sequenced the human genome and it's all stored on a massive server somewhere. Isn't it now just a question of interpretation and data-crunching? In fact, although today's version is a lot more accurate than the first draft published in 2001, the human genome is still very much a work in progress. Researchers now talk about assembling the 'platinum genome' – the one true human recipe book – while the rest of us mere mortals carry our own peculiar personal version. This also means sifting through all the hard-to-read junk, rubbish and garbage DNA, to check that there's nothing important lurking in there before it gets dismissed. As Ewan Birney (he of the human genome sweepstake) puts it, 'It's like mapping Europe and somebody says, "Oh, there's Norway. I really don't want to have to do the fjords." Now somebody's in there and mapping the fjords.'

Despite this, there's no such thing as the perfect human genome in reality. You have yours, I have mine, everybody else has theirs. And it's not the case that the DNA sequence is exactly the same in every cell of your body, no matter what the textbooks may say. That's not to say that efforts to figure out the 'über-genome' are a waste of time. Once we have a better idea of what's meant to be in our genome, it makes it easier to spot how things might have changed in individuals, tissues, tumours, diseases and even single cells.* The current way of tackling this challenge is through sequencing in ever more detail, right down to reading the DNA and RNA letters spooling from a single cell. Although it sounds challenging, that's the relatively easy bit. The hard stuff is interpreting all the data, and figuring out what (if anything) these genetic variations do, and how they work together to make us who we are.

At the moment, research focused on connecting the recipes in our genes to their resulting effects on our physical form is all about statistical relationships: finding particular DNA sequences that seem to be associated with specific traits, looking more at the level of populations rather than individual people. In 1964, the geneticist J. B. S. Haldane wrote that this type of statistical analysis 'does not explain the physiological interaction of genes and the interaction of genotype and environment. If they did so they would not be a branch of biology. They would be biology.'

To really nail down how these variations affect gene activity, scientists are going to have to turn the computer off, roll their sleeves up and do some proper experiments. It's arduous work, but in the past couple of years a set of powerful new gene-editing tools, known as CRISPR and Cas, have been developed that should help to speed things up. They're

---

*To add to the complication, as well as the standard A, T, G, C of our genetic alphabet, more sensitive DNA sequencing techniques are revealing that our genomes may be shot through with non-standard letters. As well as methylated C (the epigenetic mark we met in Chapter 10) and inosine (Chapter 12), there are several other variant letters that crop up across a wide range of organisms from bacteria to humans. However, their impact – if any – on gene activity is still mostly a mystery.

based on the molecular scissors that bacteria use to protect themselves against viruses by cutting up the invading viral DNA. As with the genetic engineering revolution in the 1970s, scientists have been quick to co-opt these tools to chop out and replace specific DNA sequences in any organism with an incredible level of precision.

Using this method, it's possible to tweak specific DNA control sequences in the genome of a living organism and see what happens.* As software gets more sophisticated and the data stack up, and as researchers become more adept at carving up the genome with their editing tools, we're getting closer to Haldane's vision of finally tying everything together. Rather than phone books full of A, C, T and G, *this* is biology nowadays.

❧

My inspiration for writing this book originated with Ernest Hemingway's six-toed pets, roaming in the sunshine on their Florida estate. During my journey, I've certainly felt like I've been herding cats – grasping after slippery ideas and trying to pin down concepts that wriggle and shift as new data come to light. Things that I thought were solid fact have been exposed as dogma and scientific hearsay, based on little evidence but repeated enough times by researchers, textbooks and journalists until they feel real. Ironically, even the story of the Hemingway cats seems to have been an invention, told time and again by tour guides at the writer's home until it became the truth. According to Hemingway scholar James

---

*As I write, researchers in China have just announced that they have used these tools to alter genes in non-viable human embryos, sparking much discussion and ethical agonising over the potential for human genetic modification, 'designer babies' and more. While these tools might be useful for fixing genetic defects due to single specific mutations, I feel that the complexities of the genome mean that we're unlikely to see engineering for more complicated traits in the near future, if ever.

Nagel, the writer didn't even *have* cats when he lived at Key West, as his wife wanted peacocks instead.

If you were hoping for a nice, neat conclusion to this book wrapping everything up and explaining how your genes work, then I'm afraid there isn't one. There's a lot we do know, and hopefully I've given you a flavour of some of the things researchers have discovered along the way. But there's a great deal that is currently unknown, and probably many things that are simply unknowable. What's clear is that we need to banish the idea that our genome is a fixed, deterministic blueprint that controls how we turn out, right from the moment when egg and sperm meet. Being alive and existing in our environment is what constructs us, in all our wobbly, unique and mysterious glory. Enjoy it.

Glossary

Amino acids: The chemical building blocks that proteins are made of.

Bases/base pairs: The chemical building blocks that DNA and RNA are made of. There are four bases, known as A (adenine), C (cytosine), G (guanine) and T (thymine) and they pair up in a consistent way – A with T, G with C. These are the 'letters' of DNA and RNA, and I've tended to use the term letters throughout to refer to them.

Chromatin: The term used to describe DNA plus its packaging proteins, known as histones.

Chromosome: A single long string of DNA. Humans have 23 pairs of chromosomes, and we get one of each pair from Mum and one from Dad.

Control switch: A stretch of DNA that turns a particular gene on or off. Scientists refer to these as regulatory elements, and switches that turn on a gene are known as enhancers.

DNA: Short for deoxyribonucleic acid, this is a long molecular string in the shape of a ladder (known as double-stranded). The outside struts are a chain of sugary molecules, while the rungs are pairs of chemicals called bases. The specific order of these bases carries genetic information which the cell uses to make the molecules it needs to stay alive and function properly.

DNA methylation: A tiny chemical tag attached to the letter C at certain points within DNA.

Enhancer: A control switch that helps to turn a gene on.

Gene: A specific stretch of DNA that carries the information for the cell to make a particular protein or RNA.

Genome: The entire set of DNA in a single cell of an organism. All our cells contain the same DNA (more or less), so the genome in a skin cell should be the same as that in a liver cell from the same person, for example.

Genotype: The particular genetic makeup of an individual organism.

Germline/Germ cells: The special cells in an embryo that will become either eggs or sperm.

Histones: The blobby proteins that package DNA in the nucleus.

Kilobase: A thousand DNA or RNA letters.

Megabase: A million DNA or RNA letters.

Non-coding DNA: A stretch of DNA that doesn't carry the instructions to make a protein. It might do nothing, or it might be used as a template to make non-coding RNA.

Non-coding RNA: An RNA message that is transcribed from DNA but doesn't carry the instructions to make a protein. A huge number of these non-coding RNA messages are made in cells, although it's unclear what all of them are doing.

Nucleus: The structure within a cell where all the DNA is housed, and where genes are read and RNA is made – it can be thought of as the 'control centre' of the cell. The rest of the cell is broadly referred to as the cytoplasm, where proteins are made and all sorts of other stuff goes on.

Phenotype: How an organism looks and behaves.

Promoter: The start of a gene.

Protein: A molecule made up of a long string of small building blocks called amino acids. Proteins do virtually all the work in cells, from maintaining their structure to carrying out the chemical reactions that keep us alive.

Ribosome: A molecular 'factory' that uses the instruments in RNA to build proteins (known as translation).

RNA: Short for ribonucleic acid, this is a chemical very similar to DNA and containing bases in a particular order, but it is usually only single-stranded, like a ladder that has been cut vertically down through the rungs. RNA is effectively a single-stranded copy of DNA, produced when a gene is read.

RNA polymerase: The molecular 'machine' that builds a string of RNA message from the template encoded in DNA.

Sequencing: Reading the order of letters (bases) in any stretch of DNA or RNA.

Transcription: The act of 'reading' a DNA template and making RNA from it.

Transcription factor: A special protein that helps to switch a gene on, meaning that it is read into RNA.

Translation: The act of assembling a protein by 'reading' a string of RNA, carried out by ribosomes.

Further Reading

This list covers the key research findings mentioned in each chapter, along with relevant books and web links. There are also some other suggestions for further reading if you want to explore further.

Introduction: It's All About That Base

Quote from presentation speech by Professor Sven Gard, member of the Nobel Committee for Physiology or Medicine, 1965. bit.ly/1G7PcRM

Chapter 1: It's Not What You've Got, It's What You Do With It That Counts

Kari Stefansson quote from a talk given on 2 December 2014 at The Commercialisation of Life meeting run by the Progress Educational Trust.

A. M. Maxam and W. Gilbert. 1977. A new method for sequencing DNA. *PNAS* 74: 560–4.

F. Sanger *et al.* 1977. DNA sequencing with chain-terminating inhibitors. *PNAS* 74: 5463–7.

W. Fiers *et al.* 1976. Complete nucleotide sequence of bacteriophage MS2 RNA: primary and secondary structure of the replicase gene. *Nature* 260: 500–7.

R. D. Fleischmann *et al.* 1995. Whole-genome random sequencing and assembly of *Haemophilus influenzae* Rd. *Science* 269: 496–512.

A. Goffeau *et al.* 1996. Life with 6000 genes. *Science* 274: 546, 563–7.

C. elegans Sequencing Consortium. 1998. Genome sequence of the nematode *C. elegans*: a platform for investigating biology. *Science* 282: 2012–18.

BBC News. 2000. Leaders' genetic code warning. *BBC News*, 26 June. bbc.in/1Hc3eUe

International Human Genome Sequencing Consortium. 2001. Initial sequencing and analysis of the human genome. *Nature* 409: 860–921.

C. Q. Choi. 2003. Who'll sweep the Gene Sweepstake? *Genome Biology* 4: spotlight.

H. Pearson. 2003. Human gene number wager won. *Nature* (online), 3 June. bit.ly/1DiGMeF

David Bentley quote taken from *Wellcome News Supplement 4: Unveiling the Human Genome* (Wellcome Trust, London, 2001).

I. Ezkurdia *et al.* 2014. Multiple evidence strands suggest that there may be as few as 19 000 human protein-coding genes. *Human Molecular Genetics* 23: 5866–78.

J. K. Colbourne *et al.* 2011. The ecoresponsive genome of *Daphnia pulex*. *Science* 331: 555–61.

T. H. Saey. 2010. More than a chicken, fewer than a grape. *ScienceNews*, 6 November. bit.ly/1KUmmdX

R. Velasco *et al.* 2010. The genome of the domesticated apple (*Malus* × *domestica Borkh.*) *Nature Genetics* 42: 833–9.

R. Brenchley *et al.* 2012. Analysis of the bread wheat genome using whole-genome shotgun sequencing. *Nature* 491: 705–10.

D. Graur. 2013. The Origin of the Term 'Junk DNA': A Historical Whodunnit. *Judge Starling*, 19 October. bit.ly/1Rlfs79

S. Ohno. 1972. So much 'junk' DNA in our genome. *Brookhaven Symposia in Biology* 23: 366–70. junkdna.com/ohno.html

E. Pennisi. 2012. ENCODE project writes eulogy for junk DNA. *Science* 337: 1159–61.

F. Macrae. 2012. The 'dark matter DNA' that could revolutionise medicine - even though scientists had written it off as junk. *Daily Mail*, 5 September. dailym.ai/1RlfQ5m

A. Jha. 2012. Breakthrough study overturns theory of 'junk DNA' in genome. *The Guardian*, 5 September. bit.ly/1Hc8hnv

The ENCODE Project Consortium. 2012. An integrated encyclopedia of DNA elements in the human genome. *Nature* 489: 57–74.

ENCODE: The encyclopedia of DNA elements. encodeproject.org

Chapter 2: Taking Out the Garbage

Dan Graur's Judge Starling blog. judgestarling.tumblr.com

M. R. Rampino and S. Self. 1993. Bottleneck in human evolution and the Toba eruption. *Science* 262: 1955.

C. S. Lane *et al.* 2013. Ash from the Toba supereruption in Lake Malawi shows no volcanic winter in East Africa at 75 ka. *PNAS* 110: 8025–9.

L. Edwards. 2010. Humans were once an endangered species. *Phys. Org*, 21 January. bit.ly/1Tmj5Hi

C. D. Huff *et al.* 2010. Mobile elements reveal small population size in the ancient ancestors of *Homo sapiens*. *PNAS* 107: 2147–52.

A. Dey *et al.* 2013. Molecular hyperdiversity defines populations of the nematode *Caenorhabditis brenneri*. *PNAS* 110: 11056–60.

D. Graur *et al.* 2013. On the immortality of television sets: 'function' in the human genome according to the evolution-free gospel of ENCODE. *Genome Biology and Evolution* 5: 578–90.

A variation of the J. B. S. Haldane beetle quote appears in Haldane's book *What is Life?* (Boni and Gaer, New York, 1947), although he was apparently fond of saying it all the time.

D. Graur. 2015. All Junk DNA May Become Functional: All Children Born in USA May Become Presidents. *Judge Starling*, 26 March. bit.ly/1S9E8uo

C. M. Rands *et al.* 2014. 8.2% of the human genome is constrained: variation in rates of turnover across functional element classes in the human lineage. *PLoS Genetics* 10: e1004525.

S. Brenner. 1998. Refuge of spandrels. *Current Biology* 8: R669.

D. Graur *et al.* 2015. An evolutionary classification of genomic function. *Genome Biology and Evolution* 7: 642–5.

Chapter 3: A Bit of Dogma

There are many excellent online resources to learn more about molecular biology and biochemistry including the University of Utah's Learn Genetics site (learn.genetics.utah.edu) and Nature's Scitable (bit.ly/1Cq3S3X).

F. Crick. 1970. Central dogma of molecular biology. *Nature* 227: 561–3.

H. M. Temin and S. Mizutani. 1970. Viral RNA-dependent DNA polymerase: RNA-dependent DNA polymerase in virions of Rous sarcoma virus. *Nature* 226: 1211–13.

D. Baltimore. 1970. Viral RNA-dependent DNA polymerase: RNA-dependent DNA polymerase in virions of RNA tumour viruses. *Nature* 226: 1209–11.

Chapter 4: Throwing the Switch

F. Jacob and J. Monod. 1961. Genetic regulatory mechanisms in the synthesis of proteins. *Journal of Molecular Biology* 3: 318–56.

Chapter 5: The Secret's in the Blend

M. Ptashne. 1986. *A Genetic Switch*. Cell Press, Cambridge, Massachusetts.

M. Ptashne. 2004. *A Genetic Switch, Phage Lambda Revisited,* third edition. Cold Spring Harbor Laboratory Press, New York.

M. Ptashne. 2011. Principles of a switch. *Nature Chemical Biology* 7: 484–7.

M. Ptashne. 2014. The chemistry of regulation of genes and other things. *Journal of Biological Chemistry* 289: 5417–35.

M. Ptashne and A. Gann. 1997. Transcriptional activation by recruitment. *Nature* 386: 569–77.

M. Ptashne. 2009. Binding reactions: epigenetic switches, signal transduction and cancer. *Current Biology* 19: R234–41.

X. Wang *et al.* 2011. Nucleosomes and the accessibility problem. *Trends in Genetics* 27: 487–92.

M. Ptashne and A. Gann. 2002. *Genes and Signals.* Cold Spring Harbor Laboratory Press, New York.

Chapter 6: Cats with Thumbs

C. Nüsslein-Volhard and E. Wieschaus. 1980. Mutations affecting segment number and polarity in *Drosophila*. *Nature* 287: 795–801.

S. Krauss *et al.* 1993. A functionally conserved homolog of the Drosophila segment polarity gene hh is expressed in tissues with polarizing activity in zebrafish embryos. *Cell* 75: 1431–44.

R. D. Riddle *et al.* 1993. Sonic hedgehog mediates the polarizing activity of the ZPA. *Cell* 75: 1401–16.

A. Keen and C. Tabin. 2004. Cliff Tabin: Super Sonic – An Interview. *The Weekly Murmur,* 12 April. bit.ly/1G8ojdn

J. Sharpe *et al.* 1999. Identification of Sonic hedgehog as a candidate gene responsible for the polydactylous mouse mutant Sasquatch. *Current Biology* 9: 97–100.

E. Anderson *et al.* 2014. Mapping the Shh long-range regulatory domain. *Development* 141: 3934–43.

R. E. Hill and L. A. Lettice. 2013. Alterations to the remote control of Shh gene expression cause congenital abnormalities. *Philosophical Transactions of the Royal Society of London B: Biological Sciences* 368: 20120357.

E. Anderson *et al.* 2012. Human limb abnormalities caused by disruption of hedgehog signaling. *Trends in Genetics* 28: 364–73.

L. A. Lettice *et al.* 2002. Disruption of a long-range cis-acting regulator for Shh causes preaxial polydactyly. *PNAS* 99: 7548–53.

L. A. Lettice *et al.* 2003. A long-range Shh enhancer regulates expression in the developing limb and fin and is associated with preaxial polydactyly. *Human Molecular Genetics* 12: 1725–35.

Chapter 7: Fish with Hips

T. Lamonerie *et al.* 1996. Ptx1, a bicoid-related homeo box transcription factor involved in transcription of the pro-opiomelanocortin gene. *Genes & Development* 10: 1284–95.

J. Shang *et al.* 1997. Backfoot, a novel homeobox gene, maps to human chromosome 5 (BFT) and mouse chromosome 13 (Bft). *Genomics* 40: 108–13.

M. D. Shapiro *et al.* 2004. Genetic and developmental basis of evolutionary pelvic reduction in threespine sticklebacks. *Nature* 428: 717–23.

C. T. Miller *et al.* 2007. cis-Regulatory changes in Kit ligand expression and parallel evolution of pigmentation in sticklebacks and humans. *Cell* 131: 1179–89.

C. A. Guenther *et al.* 2014. A molecular basis for classic blond hair color in Europeans. *Nature Genetics* 46: 748–52.

K. Conger. 2014. It's a blond thing: Stanford researchers suss out molecular basis of hair color. *Scope*, 2 June. http://stan.md/1MSkROs

C. Y. McLean *et al.* 2011. Human-specific loss of regulatory DNA and the evolution of human-specific traits. *Nature* 471: 216–19.

Chapter 8: Mice and Men and Mole Rats, Oh My!

M. C. King and A. C. Wilson. 1975. Evolution at two levels in humans and chimpanzees. *Science* 188: 107–16.

C. Dreifus. 2015. A Never-Ending Genetic Quest: Mary-Claire King's Pioneering Gene Work, From Breast Cancer to Human Rights. *New York Times*, 9 February. nyti.ms/1M2puFt

Y. Gilad *et al.* 2006. Expression profiling in primates reveals a rapid evolution of human transcription factors. *Nature* 440: 242–5.

D. Villar *et al.* 2015. Enhancer evolution across 20 mammalian species. *Cell* 160: 554–66.

H. Thompson. 2012. An Evolutionary Whodunit: How Did Humans Develop Lactose Tolerance? *NPR: The Salt*, 28 December. n.pr/1NTYiF8

M. Lynch. 2010. Rate, molecular spectrum, and consequences of human mutation. *PNAS* 107: 961–8.

Chapter 9: Party Town

S. Boyle *et al.* 2011. The spatial organization of human chromosomes within the nuclei of normal and emerin-mutant cells. *Human Molecular Genetics* 10: 211–19.

P. Therizols *et al.* 2014. Chromatin decondensation is sufficient to alter nuclear organization in embryonic stem cells. *Science* 346: 1238–42.

I. Williamson *et al.* 2014. Spatial genome organization: contrasting views from chromosome conformation capture and fluorescence in situ hybridization. *Genes & Development* 28: 2778–91.

E. Pennisi. 2015. Inching toward the 3D genome. *Science* 347: 10.

S. S. Rao *et al.* 2014. A 3D map of the human genome at kilobase resolution reveals principles of chromatin looping. *Cell* 159: 1665–80.

Chapter 10: Pimp My Genome

C. H. Waddington. 1942. The epigenotype. *Endeavour* 1: 18.

For a general (and somewhat overenthusiastic) overview of the field of epigenetics see N. Carey. 2012. *The Epigenetics Revolution.* Icon Books Ltd, London. For a more cynical perspective see M. Ptashne. 2013. Epigenetics: Core misconcept. *PNAS* 110: 7101–3; M. Ptashne. 2013. Faddish stuff: epigenetics and the inheritance of acquired characteristics. *FASEB Journal* 27(1): 1–2; and M. Ptashne. 2007. On the use of the word 'epigenetic'. *Current Biology* 17: R233–6.

A. Bird *et al.* 1979. Methylated and unmethylated DNA compartments in the sea urchin genome. *Cell* 17: 889–901.

J. Boyes and A. Bird. 1991. DNA methylation inhibits transcription indirectly via a methyl-CpG binding protein. *Cell* 64: 1123–34.

P. Green and A. F. Rubin. 2009. Mutation patterns in cancer genomes. *PNAS* 106: 21766–70.

S. Yi. 2012. Birds do it, bees do it, worms and ciliates do it too: DNA methylation from unexpected corners of the tree of life. *Genome Biology* 13: 174.

F. Capuano. 2014. Cytosine DNA methylation is found in *Drosophila melanogaster* but absent in *Saccharomyces cerevisiae, Schizosaccharomyces pombe,* and other yeast species. *Analytical Chemistry* 86: 3697–3702.

M. Bachman *et al.* 2014. 5-hydroxymethylcytosine is a predominantly stable DNA modification. *Nature Chemistry* 6: 1049–55.

G. Zhang *et al.* 2015. N6-methyladenine DNA modification in *Drosophila*. *Cell* 161: 893–906.

E. L. Greer *et al.* 2015. DNA methylation on N6-Adenine in *C. elegans*. *Cell* (2015) 161: 868–78.

Y. Fu *et al.* 2015. N6-methyldeoxyadenosine marks active transcription start sites in *Chlamydomonas*. *Cell* 161: 879–92.

J. Greally. 2015. Human Disease Epigenomics 2.0. *PLoS Biologue*, 7 July. bit.ly/1fAFmmr

P. N. Audergon *et al.* 2015. Restricted epigenetic inheritance of H3K9 methylation. *Science* 348: 132–5.

J. A. Hackett *et al.* 2012. Promoter DNA methylation couples genome-defence mechanisms to epigenetic reprogramming in the mouse germline. *Development* 139: 3623–32.

D. Roulois et al. 2015. DNA-demethylating agents target colorectal cancer cells by inducing viral mimicry by endogenous transcripts. *Cell* 162: 961–973.

B. M. Nugent *et al.* 2015. Brain feminization requires active repression of masculinization via DNA methylation. *Nature Neuroscience* 18: 690–7.

J. B. Gurdon. 1962. The developmental capacity of nuclei taken from intestinal epithelium cells of feeding tadpoles. *Journal of Embryology and Experimental Morphology* 34: 93–112.

I. Wilmut *et al.* 1997. Viable offspring derived from fetal and adult mammalian cells. *Nature* 385: 810–13.

Wikipedia: List of animals that have been cloned. bit.ly/1MkQj5P

K. Takahashi and S. Yamanaka. 2006. Induction of pluripotent stem cells from mouse embryonic and adult fibroblast cultures by defined factors. *Cell* 126: 663–76.

I. C. G. Weaver *et al.* 2004. Epigenetic programming by maternal behavior. *Nature Neuroscience* 7: 847–54.

P. O. McGowan *et al.* 2009. Epigenetic regulation of the glucocorticoid receptor in human brain associates with childhood abuse. *Nature Neuroscience* 12: 342–8.

V. Hughes. 2013. Mice Inherit Specific Memories, Because Epigenetics? *Phenomena: Only Human*, 1 December. bit.ly/1Cu1Pfd

R. Yehuda *et al.* 2005. Transgenerational effects of posttraumatic stress disorder in babies of mothers exposed to the World Trade Center attacks during pregnancy. *Journal of Clinical Endocrinology and Metabolism* 90: 4115–18.

The Guardian. 2011. Pregnant 9/11 survivors transmitted trauma to their children. *The Guardian*, 9 September. bit.ly/1eNtYCI

E. Pilkington. 2014. Tennessee set to criminalise pregnant women who use illegal drugs. *The Guardian*, 19 April. bit.ly/1RoMrrc

L. R. McRobbie. 2012. 12 Terrible Pieces of Advice for Pregnant Women. *Mental Floss*, 12 December. bit.ly/1Mo3VXg

M. Jeay and K. Garay. 2006. *The Distaff Gospels*. Broadview Editions, Peterborough, Ontario.

Chapter 11: Cut and Paste

H. C. Friedmann. 2004. From '*Butyribacterium*' to '*E. coli*': an essay on unity in biochemistry. *Perspectives in Biology and Medicine* 47: 47–66.

G. Cooper. 1996. Bums, biceps and Bunsen burners. *The Independent*, 9 January. ind.pn/1dWWIZ4

L. T. Chow *et al*. 1977. An amazing sequence arrangement at the 5′ ends of adenovirus 2 messenger RNA. *Cell* 12: 1–8.

J. M. Adams and S. Cory. 1975. Modified nucleosides and bizarre 5′-termini in mouse myeloma mRNA. *Nature* 255: 28–33.

S. M. Berget *et al*. 1977. Spliced segments at the 5′ terminus of adenovirus 2 late mRNA. *PNAS* 74: 3171–5.

J. Clancy. Stonesoup cookery blog. thestonesoup.com

D. Schmucker *et al*. 2000. *Drosophila* Dscam is an axon guidance receptor exhibiting extraordinary molecular diversity. *Cell* 101: 671–84.

Chapter 12: Nature's Red Pen

ALS Association. 2014. The ALS Association Expresses Sincere Gratitude to Over Three Million Donors. *ALS Association*, 29 August. bit.ly/1Cvc9mZ

R. Dalton and Q. Schiermeier. 1999. Genentech pays $200m over growth hormone 'theft'. *Nature* 402: 335.

B. Sommer *et al*. 1991. RNA editing in brain controls a determinant of ion flow in glutamate-gated channels. *Cell* 67: 11–19.

R. Brusa *et al*. 1995. Early-onset epilepsy and postnatal lethality associated with an editing-deficient GluR-B allele in mice. *Science* 270: 1677–80.

Y. Kawahara *et al*. 2004. Glutamate receptors: RNA editing and death of motor neurons. *Nature* 427: 801.

T. Yamashita *et al.* 2013. Rescue of amyotrophic lateral sclerosis phenotype in a mouse model by intravenous AAV9-ADAR2 delivery to motor neurons. *EMBO Molecular Medicine* 5: 1710–19.

J. B. Li *et al.* 2009. Genome-wide identification of human RNA editing sites by parallel DNA capturing and sequencing. *Science* 324: 1210–13.

Chapter 13: Ever-increasing Circles

J. Wapner. 2014. *The Philadelphia Chromosome: A Genetic Mystery, a Lethal Cancer, and the Improbable Invention of a Lifesaving Treatment.* The Experiment LLC, New York.

J. Salzman *et al.* 2012. Circular RNAs are the predominant transcript isoform from hundreds of human genes in diverse cell types. *PLoS One* 7: e30733.

B. Capel *et al.* 1993. Circular transcripts of the testis-determining gene Sry in adult mouse testis. *Cell* 73: 1019–30.

J. M. Houseley. 2006. Noncanonical RNAs from transcripts of the *Drosophila* muscleblind gene. *Journal of Heredity* 97: 253–60.

X. You *et al.* 2015. Neural circular RNAs are derived from synaptic genes and regulated by development and plasticity. *Nature Neuroscience* 18: 603–10.

M. W. Hentze and T. Preiss. 2013. Circular RNAs: splicing's enigma variations. *EMBO Journal* 32: 923–5.

Chapter 14: Silence of the Genes

G. L. Sen and H. M. Blau. 2006. A brief history of RNAi: the silence of the genes. *FASEB Journal* 20: 1293–9.

C. Napoli *et al.* 1990. Introduction of a chimeric chalcone synthase gene into petunia results in reversible co-suppression of homologous genes in trans. *Plant Cell* 2: 279–89.

N. Romano and G. Macino. 1992. Quelling: transient inactivation of gene expression in *Neurospora crassa* by transformation with homologous sequences. *Molecular Microbiology* 6: 3343–53.

S. Guo and K. Kemphues. 1995. Par-1, a gene required for establishing polarity in *C. elegans* embryos, encodes a putative Ser/Thr kinase that is asymmetrically distributed. *Cell* 81: 611–20.

A. Fire *et al.* 1998. Potent and specific genetic interference by double-stranded RNA in *Caenorhabditis elegans*. *Nature* 391: 806–11.

J. Ruttiman. 2006. RNA therapy tackles eye disease. *Nature News*, 6 June. bit.ly/1gsNZzz

A. Maxmen. 2013. Mammals chop up viral RNA to attack infection. *Nature News*, 10 October. bit.ly/1J9mlhU

R. C. Lee *et al.* 1993. The *C. elegans* heterochronic gene lin-4 encodes small RNAs with antisense complementarity to lin-14. *Cell* 75: 843–54.

E. Londin *et al.* 2015. Analysis of 13 cell types reveals evidence for the expression of numerous novel primate- and tissue-specific microRNAs. *PNAS* 112: E1106–15.

Y. Lee *et al.* 2004. MicroRNA genes are transcribed by RNA polymerase II. *EMBO Journal* 23: 4051–60.

J. A. Vidigal and A. Venture. 2015. The biological functions of miRNAs: lessons from *in vivo* studies. *Trends in Cell Biology* 25: 137–47.

E. A. Miska. 2007. Most *Caenorhabditis elegans* microRNAs are individually not essential for development or viability. *PLoS Genetics* 3: e215.

M. Baker. 2009. Haifan Lin: peeling back layers of stem cell control. *Nature Reports Stem Cells*, 12 February. bit.ly/1CvkJCg

E-M. Weick and E. A. Miska. 2014. piRNAs: from biogenesis to function. *Development* 141: 3458–71.

E. Ladoukakis *et al.* 2011. Hundreds of putatively functional small open reading frames in *Drosophila*. *Genome Biology* 12: R118.

A. A. Bazzini *et al.* 2014. Identification of small ORFs in vertebrates using ribosome footprinting and evolutionary conservation. *EMBO Journal* 33: 981–93.

Chapter 15: Night of the Living Dead

P. M. Harrison *et al.* 2002. Molecular fossils in the human genome: identification and analysis of the pseudogenes in chromosomes 21 and 22. *Genome Research* 12: 272–80.

R. Sasidharan and M. Gerstein. 2008. Genomics: Protein fossils live on as RNA. *Nature* 453: 729–31.

N. A. Rapicavoli *et al.* 2013. A mammalian pseudogene lncRNA at the interface of inflammation and anti-inflammatory therapeutics. *eLife* 2: e00762.

Chapter 16: On the Hop

D. M. Sayah *et al.* 2004. Cyclophilin A retrotransposition into TRIM5 explains owl monkey resistance to HIV-1. *Nature* 430: 569–73.

International Human Genome Sequencing Consortium. 2001. Initial sequencing and analysis of the human genome. *Nature* 409: 860–921.

E. Fox Keller. 1983. *A Feeling for the Organism: The Life and Work of Barbara McClintock*. W. H. Freeman and Company, San Francisco, California.

F. S. de Souza *et al.* 2013. Exaptation of transposable elements into novel cis-regulatory elements: is the evidence always strong? *Molecular Biology and Evolution* 30: 1239–51.

K. R. Upton. 2011. Is somatic retrotransposition a parasitic or symbiotic phenomenon? *Mobile Genetic Elements* 1: 279–82.

A. D. Ewing *et al.* 2013. Retrotransposition of gene transcripts leads to structural variation in mammalian genomes. *Genome Biology* 14: R22.

A. R. Muotri *et al.* 2005. Somatic mosaicism in neuronal precursor cells mediated by L1 retrotransposition. *Nature* 435: 903–10.

N. G. Coufal *et al.* 2009. L1 retrotransposition in human neural progenitor cells. *Nature* 460: 1127–31.

J. K. Baillie *et al.* 2011, Somatic retrotransposition alters the genetic landscape of the human brain. *Nature* 479: 534–7.

G. D. Evrony *et al.* 2012. Single-neuron sequencing analysis of L1 retrotransposition and somatic mutation in the human brain. *Cell* 151: 483–96.

M. Bundo *et al.* 2014. Increased L1 retrotransposition in the neuronal genome in schizophrenia. *Neuron* 81: 306–13.

R. C. Iskow *et al.* 2010. Natural mutagenesis of human genomes by endogenous retrotransposons. *Cell* 141: 1253–61.

Y. Miki *et al.* 1992. Disruption of the APC gene by a retrotransposal insertion of L1 sequence in a colon cancer. *Cancer Research* 52: 643–5.

E. Lee *et al.* 2012. Landscape of somatic retrotransposition in human cancers. *Science* 337: 967–71.

A. L. Paterson *et al.* 2015. Mobile element insertions are frequent in oesophageal adenocarcinomas and can mislead paired-end sequencing analysis. *BMC Genomics* 16: 473.

E. J. Grow *et al.* 2015. Intrinsic retroviral reactivation in human preimplantation embryos and pluripotent cells. *Nature* 522: 221–5.

K. Conger. 2015. Viral proteins may regulate human embryonic development. *Phys.Org*, 21 April. bit.ly/1HksoSb

Chapter 17: Opening a Can of Wobbly Worms

M. Francesconi and B. Lehner. 2014. The effects of genetic variation on gene expression dynamics during development. *Nature* 505: 208–11.

A. Burga *et al.* 2011. Predicting mutation outcome from early stochastic variation in genetic interaction partners. *Nature* 480: 250–3.

M. O. Casanueva *et al.* 2012. Fitness trade-offs and environmentally induced mutation buffering in isogenic *C. elegans*. *Science* 335: 82–5.

N. Frankel *et al.* 2010. Phenotypic robustness conferred by apparently redundant transcriptional enhancers. *Nature* 466: 490–3.

Phys.Org. 2015. 'Quantum jitters' could form basis of evolution, cancer. *Phys.Org*, 11 March. bit.ly/1Mor1gc

A. J. Kimsey. 2015. Visualizing transient Watson–Crick-like mispairs in DNA and RNA duplexes. *Nature* 519: 315–20.

Chapter 18: Everyone's a Little Bit Mutant

A. F. Rope *et al.* 2011. Using VAAST to identify an X-linked disorder resulting in lethality in male infants due to N-terminal acetyltransferase deficiency. *American Journal of Human Genetics* 89: 28–43.

L. C. Francioli *et al.* 2014. Whole-genome sequence variation, population structure and demographic history of the Dutch population. *Nature Genetics* 46: 818–25.

D. S. Herman *et al.* 2012. Truncations of titin causing dilated cardiomyopathy. *New England Journal of Medicine* 366: 619–28.

A. M. Roberts *et al.* 2015. Integrated allelic, transcriptional, and phenomic dissection of the cardiac effects of titin truncations in health and disease. *Science Translational Medicine* 7: 270ra6.

E. C. Hayden. 2015. Pint-sized DNA sequencer impresses first users. *Nature News*, 5 May. bit.ly/1gsStq7

Chapter 19: Opening the Black Box

Mike Dexter quote from *Wellcome News Supplement 4: Unveiling the Human Genome* (Wellcome Trust, London, 2001).

D. L. Hart and D. J. Fairbanks. 2007. Mud sticks: On the alleged falsification of Mendel's data. *Genetics* 175: 975–9.

E. Novitski. 2004. On Fisher's criticism of Mendel's results with the garden pea. *Genetics* 166: 1133–6.

Figure Concord: Mendel's Bees. bit.ly/1HTlTrC

Weldon's letters cited in *A Century of Mendelism in Human Genetics*, edited by M. Keynes, A. W. F. Edwards and R. Peel (CRC Press, Boca Raton, Florida, 2004).

A. Jamieson and G. Radick. 2013. Putting Mendel in his place: how curriculum reform in genetics and counterfactual history of science can work together. In: *The Philosophy of Biology: A Companion for Educators*, ed. K. Kampourakis. Springer, Dordrecht, Netherlands.

G. Radick. 2005. Other histories, other biologies. In: *Philosophy, Biology and Life,* ed. Anthony O'Hear. Cambridge University Press, Cambridge, UK.

W. F. R. Weldon. 1902. Mendel's laws of alternative inheritance in peas. *Biometrika* 1: 228–54.

Cancer Research UK. 2014. Smartphone gamers decode half a year of Cancer Research UK's genetic data in one month. *Cancer Research UK press release*, 14 March. bit.ly/1CvvUuA

R. Chen *et al.* 2012. Personal omics profiling reveals dynamic molecular and medical phenotypes. *Cell* 148: 1293–1307.

The 'Snyderome'. snyderome.stanford.edu

Chapter 20: Blame the Parents

M. H. Kaufman *et al.* 1977. Normal postimplantation development of mouse parthenogenetic embryos to the forelimb bud stage. *Nature* 265: 53–5.

M. A. Surani *et al.* 1977. Development to term of chimaeras between diploid parthenogenetic and fertilised embryos. *Nature* 270: 601–3.

M. A. Surani *et al.* 1984. Development of reconstituted mouse eggs suggests imprinting of the genome during gametogenesis. *Nature* 308: 548–50.

S. C. Barton *et al.* 1984. Role of paternal and maternal genomes in mouse development. *Nature* 311: 374–6.

M. A. Surani *et al.* 1986. Nuclear transplantation in the mouse: heritable differences between parental genomes after activation of the embryonic genome. *Cell* 45: 127–36.

J. McGrath and D. Solter. 1984. Completion of mouse embryogenesis requires both the maternal and paternal genomes. *Cell* 37: 179–83.

M. Dolasia. 2010. Researchers Disguise As Pandas To Save The Endangered Animals. *Dogo News*, 10 December. bit.ly/1eO4p4C

W. Reik *et al.* 1987. Genomic imprinting determines methylation of parental alleles in transgenic mice. *Nature* 328: 248–51.

N. D. Allen *et al.* 1988. Transgenes as probes for active chromosomal domains in mouse development. *Nature* 333: 852–5.

A. G. Searle and C. V. Beechey. 1990. Genome imprinting phenomena on mouse chromosome 7. *Genetics Research* 56: 237–44.

A. C. Ferguson-Smith *et al.* 1991. Embryological and molecular investigations of parental imprinting on mouse chromosome 7. *Nature* 351: 667–70.

P. B. Vrana *et al.* 1998. Genomic imprinting is disrupted in interspecific *Peromyscus* hybrids. *Nature Genetics* 20: 362–5.

M. A. Cleaton *et al.* 2014. Phenotypic outcomes of imprinted gene models in mice: elucidation of pre- and postnatal functions of imprinted genes. *Annual Review of Genomics and Human Genetics* 15: 93–126.

M. Charalambous *et al.* 2012. Imprinted gene dosage is critical for the transition to independent life. *Cell Metabolism* 15: 209–21.

S. R. Ferron *et al.* 2011. Postnatal loss of Dlk1 imprinting in stem cells and niche astrocytes regulates neurogenesis. *Nature* 475: 381–5.

S. T. de Rocha *et al.* 2009. Gene dosage effects of the imprinted delta-like homologue 1 (Dlk1/Pref1) in development: implications for the evolution of imprinting. *PLoS Genetics* 5: e1000392.

Chapter 21: Meet the Mickey Mouse Mice

Wikiquote: Louis Pasteur quote from a lecture given at the University of Lille, 7 December 1854. bit.ly/1gsTcaL

M. Rassoulzadegan *et al.* 2006. RNA-mediated non-mendelian inheritance of an epigenetic change in the mouse. *Nature* 441: 469–74.

V. Grandjean *et al.* 2009. The miR-124-Sox9 paramutation: RNA-mediated epigenetic control of embryonic and adult growth. *Development* 136: 3647–55.

K. D. Wagner *et al.* 2008. RNA induction and inheritance of epigenetic cardiac hypertrophy in the mouse. *Developmental Cell* 14(6): 962–9.

M. Kawano *et al.* 2012. Novel small noncoding RNAs in mouse spermatozoa, zygotes and early embryos. *PLoS One* 7: e44542.

R. Liebers *et al.* 2014. Epigenetic regulation by heritable RNA. *PLoS Genetics* 10: e1004296.

K. Gapp. *et al.* 2014. Implication of sperm RNAs in transgenerational inheritance of the effects of early trauma in mice. *Nature Neuroscience* 17: 667–9.

C. Darwin. 1868. *The Variation of Animals and Plants Under Domestication.* John Murray, London.

C. Cossetti *et al.* 2014. Soma-to-germline transmission of RNA in mice xenografted with human tumour cells: possible transport by exosomes. *PLoS One* 9: e101629.

O. J. Rando. tRNA fragments as transgenerational information carriers. NIH Director's Pioneer Award. bit.ly/1J9ryXd

Chapter 22: In Search of the 21st-Century Gene

P. Griffiths and K. Stotz. 2013. *Genetics and Philosophy.* Cambridge University Press, Cambridge, UK.

The Representing Genes Project. representinggenes.org

K. Stotz *et al.* 2004. How biologists conceptualize genes: An empirical study. *Studies in History and Philosophy of Biological and Biomedical Sciences* 35: 647–73.

R. Alleyne. 2010. Scientist Craig Venter creates life for first time in laboratory sparking debate about 'playing god'. *Daily Telegraph*, 20 May. bit.ly/1IWozpn

E. Callaway. 2014. 'Platinum' genome takes on disease. *Nature News*, 18 November. bit.ly/1G7ZSQg

J. B. S. Haldane. 2008. A defense of beanbag genetics. Reprinted in *International Journal of Epidemiology* 37: 435–42.

Y. Guo *et al.* 2015. CRISPR inversion of CTCF sites alters genome topology and enhancer/promoter function. *Cell* 162: 900–10.

J. Allen. 1999. Hemingway's Key West home a mix of writer's life, legend. *CNN.com.* bit.ly/1KUDodo

D. Cyranoski and A. Reardon. 2015. Chinese scientists genetically modify human embryos. *Nature News*, 22 April. bit.ly/1dQBsnF

P. Liang *et al.* 2015. CRISPR/Cas9-mediated gene editing in human tripronuclear zygotes. *Protein & Cell* 6: 363–72.

Further Reading

One of the best popular science books I've ever read about genetics is Lone Frank's *My Beautiful Genome* (Oneworld Publications, Oxford, UK, 2012), tackling the complexities of modern genomics with a personal angle.

If you can track it down, François Jacob's *The Possible and the Actual* (University of Washington Press, Seattle, 1982) is a wonderful read. Adapted from a series of lectures, it's thought-provoking, provocative, beautifully written and remarkably prescient.

Evelyn Fox Keller's *The Century of the Gene* (Harvard University Press, Cambridge, Massachusetts, 2000) is an elegant exploration of the scientific and philosophical aspects of genes and genomes, closing a century in which genes transformed from Mendelian to molecular.

The original and best, Charles Darwin's *Origin of Species* (reprinted by Wordsworth Classics, Ware, Hertfordshire, 1998) is really worth a read. Less technical and more readable than you might expect, it's a humble and devastatingly clear exposition of his ideas and what they mean.

In Pursuit of the Gene by James Schwartz (Harvard University Press, Cambridge, Massachusetts, 2008) is an excellent and impressively researched history of genetics from 'Darwin to DNA', as the tagline has it, including a detailed discussion of gemmules.

The story of the discovery of the structure of DNA, along with much of the rest of the history of molecular biology, is told in Horace Freeland Judson's *Eighth Day of Creation* (second edition, Penguin Books, London, 1995). There's also James Watson's personal (and somewhat biased) account, *The Double Helix* (Signet Books, New York, 1968).

The race to crack the genetic code is laid out with breathless pace by Matthew Cobb in his book *Life's Greatest Secret* (Profile Books, London, 2015), providing a deeper insight into the science, the characters and the serendipity that enabled us to read the language of life.

Matt Ridley's *Genome* (Fourth Estate, London, 1999) is a good introduction to genetics, although quite outdated now. For a more in-depth and detailed (and also a little outdated) discussion of the science of genomes and genomics, see Barry Barnes and John Dupré's *Genomes and What to Make of Them* (University of Chicago Press, Chicago, 2008).

A few more to mention if you're still interested: Armand LeRoi's *Mutants* (Harper Perennial, London, 2005); *Creation* by Adam Rutherford (Viking, London, 2013); Alice Roberts's *The Incredible Unlikeliness of Being* (Heron Books, London, 2014); and *p53: The gene that cracked the cancer code* by Sue Armstrong (Bloomsbury Sigma, London, 2014).

I'd Like to Thank ...

Although it's been my fingers on the keys, I couldn't have written this book without a lot of other people. This is the bit where I get all emotional and thank everybody who helped along the way. I'm sure I'm going to miss someone out, but here goes.

Firstly, a big thank you to the team at Bloomsbury. Special thanks to my editor Jim Martin for believing that a 'book about how your genes work' was even a vaguely good idea and supporting me throughout the process, from our first meeting over pizza and Chardonnay to the late-night editing crises and Twitter rants. Thanks to Cravendale Milk and Edward Monkton for permission to use their quotes.

This book would have been impossible without the large number of scientists who agreed to talk to me and were so generous with their time and thoughts. There's also a big supporting cast of people who have helped to shape my thinking along the way, whether I was interviewing them, chatting at scientific conferences or randomly burbling at them in the pub. I've completely lost track of how many people I've pestered with the opening line 'I'm writing a book about how genes work ...' so apologies if I've missed anyone.

Thanks to: Adam Auton, Adrian Bird, Andy Fire, Anna Middleton, Anne Brunet, Anne Ferguson-Smith, Aylwyn Scally, Azim Surani, Ben Lehner, Billy Li, Bob Hill, Carlos Bustamante, Dan Durocher, Dan Graur, Duncan Odom, Duncan Sproul, Eric Miska, Evelyn Fox Keller, François Cuzin, Gerard Evan, Gholson Lyon, Howard Chang, Jackie Dresch, James Sharpe, James Ware, Joanna Wysocka, Joe Pickrell, John Greally, Julia Salzman, Karola Stotz, Mandy Fisher, Martin Taylor, Mark Ptashne, Minoo Rassoulzadegan, Oliver Rando, Phil Sharp, Rich Roberts, Richard Meehan, Rick Young, Rob Drewell, Robin Allshire, Scott Waddell, Stephen Montgomery, Stuart Firestein, Tony Kouzarides, Tuuli Lappalainen and Wendy Bickmore.

Thanks to the Library Spider and Áine McCarthy for helping me out with access to the literature, and to Greg Radick, Annie Jamieson and Dominic Berry from the Centre for History and Philosophy of Science at the University of Leeds for

encouraging me to attend the Nurturing Genetics conference and expanding my thinking in new ways.

I'd also like to single out a few of my teachers for particular acknowledgement – I wouldn't be the scientist or communicator that I am today without their encouragement: Miss Hatcher at Marsworth Infant School, who just put me in the book corner and let me get on with it; my science teachers at the Sir Henry Floyd – Mrs Rowland-Davies, Dr Richards and Mr Harper; and Dr James Keeler at Selwyn College, Cambridge. And I'm also indebted to the researchers whose labs I've worked in – Graham Cook, Azim Surani and Mandy Fisher – for their patient mentoring and tolerance of all the broken glassware.

Travelling round the world chatting to scientists doesn't come cheap, so I'm hugely grateful to everyone who's given me a spare bed or bought me drinks and/or dinner along the way. I'm not going to try and account for every pie or pint, but special thanks to the people who put me up while I was in the United States: Abe and Anneke Sharp, Rob and Jackie Drewell and the Dresch clan, Ruth Williams and Afton Grant, and Rob and Linzi Brotherton. Much gratitude goes to Rob Drewell, who finally gave in to my cheeky blagging and managed to find some money for a transatlantic flight.

Next, thank you so much to all the friends, family, random internet acquaintances and scientists who volunteered to read the various drafts of this book along the way. It's not an easy job and I'm grateful for every single piece of feedback. You all helped to make it better, and I take full responsibility for what's left. Huge thanks to: Amy and Alistair Twiname, Anne Ferguson-Smith, Helen Arney, Lucy Durocher, Mark Stevenson, Matthew Cobb, Matthew Hunter, Rob Leworthy, Ruth Welters, Vicky Tustian and Wendy Bickmore. And a huge thank you to Liz Drewitt at Nature Edit for the eagle-eyed copy-editing.

I'm incredibly lucky to have a fantastic day job working with some of the best people you could ever hope to meet. A big shout out is due to my colleagues at Cancer Research UK for all their support, ideas, cakes and ridiculous email threads. Particular thanks are due to my managers, Nell Barrie and Steve Palmer, for allowing me to take unpaid leave to get the bulk of the writing done and for being so incredibly tolerant of my whims.

Thanks also to Chris Smith and the team at the Naked Scientists for more than a decade of science broadcasting fun; to

the Genetics Society executive committee for giving me the opportunity to report back from the frontiers of genetics by supporting my monthly Naked Genetics podcasts; to Mun-Keat Looi at *Mosaic* magazine for a valuable lesson in editing; to the SMeRFWIG crew and the London sci-comm mafia for professional development and personal friendships; Kate Dean and the Mutual Fanclub for being there right at the start, encouraging me to follow my creative dreams; the ever-fabulous Viv Parry; and to Mark Stevenson – my favourite champagne Optimist.

On to the personal stuff. Music is the other passion in my life, aside from writing and science, and creating, recording and performing with my bands Talk in Colour and Sunday Driver has kept me sane during the book-writing process. Thank you for the music: Chris, Nick, Mary, Dave and Rob (Talk in Colour), and Chandy, Joel, Mel, Chemise, Scott and Pete from Sunday Driver. Also a special mention to Andy Dobson and Samy Bishai from Digitonal.

Huge thanks to everyone on the Blue forum for all the online support, jokes and distractions – I'm incredibly lucky to be part of such a lovely corner of the interweb, and you're all awesome. *High fives* and lovely hugs to Team WineGym. And apologies to all the people to whom I've been a terrible, terrible friend while I've been down the book hole, particularly Ruth, Mazz, Vicky, Emily and Liz (Bird!). For all the catchups, gigs, parties and events I missed, and all the times I did manage to turn up but just banged on endlessly about The Book. I can come out and play now, and I promise I'll try to talk about something else.

Finally, thanks to my family for the endless support and cheerleading: Mum, Dad and Ferdy the dog; Lucy, Dan and Chloë Durocher; Helen Arney and Rob Leworthy; and Brendan and Julie Carr. And huge thanks to Ricky – I couldn't have done this without you. Thank you for the endless cups of coffee and surprise chocolate bars, the philosophising, musing and metaphors, and everything else. Let's go to the pub.

Index